FOREWORD

AC 00-45E, Aviation Weather Services, is published jointly by the Federal Aviation Administration and the National Weather Service (NWS). This document supplements the companion manual AC 00-6A, Aviation Weather, that deals with weather theories and hazards.

This advisory circular, AC 00-45E, explains weather service in general and the details of interpreting and using coded weather reports, forecasts, and observed and prognostic weather charts. Many charts and tables apply directly to flight planning and inflight decisions. It can also be used as a source of study for pilot certification examinations.

The AC 00-45E was written primarily by Kathleen Schlachter with contributions from Jon Osterberg, Doug Streu, and Robert Prentice. A special thanks to Sue Roe for her help and patience in editing this manual.

Comments and suggestions for improving this publication are encouraged and should be directed to:

National Weather Service Coordinator, W/SR64
Federal Aviation Administration
Mike Monroney Aeronautical Center
P.O. Box 25082
Oklahoma City, OK 73125-0082

Advisory Circular AC 00-45E supersedes AC 00-45D, Aviation Weather Services, revised 1995.

TABLE OF CONTENTS

Section 1 - THE AVIATION WEATHER SERVICE PROGRAM

Section 2 - AVIATION ROUTINE WEATHER REPORT (METAR)

Section 3 - PILOT AND RADAR REPORTS, SATELLITE PICTURES, AND RADIOSONDE ADDITIONAL DATA (RADATs)

Section 4 - AVIATION WEATHER FORECASTS

Section 5 - SURFACE ANALYSIS CHART

Section 6 - WEATHER DEPICTION CHART

Section 7 - RADAR SUMMARY CHART

Section 8 - CONSTANT PRESSURE ANALYSIS CHARTS

Section 9 - COMPOSITE MOISTURE STABILITY CHART

Section 10 - WINDS AND TEMPERATURES ALOFT CHART

SECTION 11 - SIGNIFICANT WEATHER PROGNOSTIC CHARTS

Section 12 - CONVECTIVE OUTLOOK CHART

Section 13 – VOLCANIC ASH ADVISORY CENTER PRODUCTS

Section 14 - TURBULENCE LOCATIONS, CONVERSION AND DENSITY ALTITUDE TABLES, CONTRACTIONS AND ACRONYMS, SCHEDULE OF PRODUCTS, NATIONAL WEATHER SERVICE STATION IDENTIFIERS, WSR-88D SITES, AND INTERNET ADDRESSES

Section 1
THE AVIATION WEATHER SERVICE PROGRAM

Providing weather service to aviation is a joint effort of the National Weather Service (NWS), the Federal Aviation Administration (FAA), the Department of Defense (DOD), and other aviation-oriented groups and individuals. This section discusses the civilian agencies of the U.S. Government and their observation and communication services to the aviation community.

NATIONAL OCEANIC AND ATMOSPHERIC ADMINISTRATION (NOAA)

The National Oceanic and Atmospheric Administration (NOAA) is an agency of the Department of Commerce. NOAA is one of the leading scientific agencies in the U.S. Government. Among its six major divisions are the National Environmental Satellite Data and Information Service (NESDIS) and the NWS.

NATIONAL ENVIRONMENTAL SATELLITE DATA AND INFORMATION SERVICE (NESDIS)

The National Environmental Satellite Data and Information Service (NESDIS) is located in Washington, D.C., and directs the weather satellite program. Figures 3-2 and 3-3 are examples of Geostationary Operational Environmental Satellite (GOES) images. These images are available to NWS meteorologists and a wide range of other users for operational use.

Satellite Analysis Branch (SAB)

The Satellite Analysis Branch (SAB) coordinates the satellite and other known information for the NOAA Volcanic Hazards Alert program under an agreement with the FAA. SAB works with the NWS as part of the Washington D.C. Volcanic Ash Advisory Center (VAAC).

NATIONAL WEATHER SERVICE (NWS)

The National Weather Service (NWS) collects and analyzes meteorological and hydrological data and subsequently prepares forecasts on a national, hemispheric, and global scale. The following is a description of the NWS facilities tasked with these duties.

National Centers for Environmental Prediction (NCEP)

There are nine separate national centers under National Centers for Environmental Prediction (NCEP), each with its own specific mission. They are the Climate Prediction Center, Space Environment Center, Marine Prediction Center, Hydrometeorological Prediction Center, Environmental Modeling Center, NCEP Center Operations, Storm Prediction Center, Aviation Weather Center, and the Tropical Prediction Center.

National Center Operations (NCO)

Located in Washington, D.C., the National Center Operations (NCO) is the focal point of the NWS's weather processing system. From worldwide weather reports, NCO prepares automated weather analysis charts and guidance forecasts for use by NWS offices and other users.

Some NCO products are specifically prepared for aviation, such as the winds and temperatures aloft forecast. Figure 4-9 is the network of forecast winds and temperatures aloft for the contiguous 48 states. Figure 4-10 shows the Alaskan and Hawaiian network.

NCO is part of VAAC, which runs an ash dispersion model. NCO works with SAB to fulfill the VAAC responsibilities to the aviation communities regarding potential volcanic ash hazards to aviation.

Storm Prediction Center (SPC)

The Storm Prediction Center (SPC) is charged with monitoring and forecasting severe weather over the 48 continental United States. Its products include convective outlooks and forecasts, as well as severe weather watches. The center also develops severe weather forecasting techniques and conducts research. The SPC is located in Norman, Oklahoma, near the heart of the area most frequently affected by severe thunderstorms.

Hydrometeorological Prediction Center (HPC)

The Hydrometeorological Prediction Center (HPC) prepares weather charts which include basic weather elements of temperature, fronts and pressure patterns.

Aviation Weather Center (AWC)

The Aviation Weather Center (AWC), located in Kansas City, Missouri, issues warnings, forecasts, and analyses of hazardous weather for aviation interests. The center identifies existing or imminent weather hazards to aircraft in flight and creates warnings for transmission to the aviation community. It also produces operational forecasts of weather conditions expected during the next 2 days that will affect domestic and international aviation interests. As a Meteorological Watch Office (MWO) under regulations of the International Civil Aviation Organization (ICAO), meteorologists in this unit prepare and issue aviation area forecasts (FAs) and inflight weather advisories (Airman's Meteorological Information [AIRMET], Significant Meteorological Information [SIGMET], and Convective SIGMETs) for the contiguous 48 states.

Tropical Prediction Center (TPC)

The Tropical Prediction Center (TPC) is located in Miami, Florida. The National Hurricane Center, as an integral part of TPC, issues hurricane advisories for the Atlantic, the Caribbean, the Gulf of Mexico, the eastern Pacific, and adjacent land areas. The center also develops hurricane forecasting techniques and conducts hurricane research. The Central Pacific Hurricane Center in Honolulu, Hawaii, issues advisories for the central Pacific Ocean.

TPC prepares and distributes tropical weather, aviation and marine analyses, forecasts, and warnings. As an MWO, TPC meteorologists prepare and issue aviation forecasts, SIGMETs, and Convective SIGMETs for their tropical Flight Information Region (FIR).

Weather Forecast Office (WFO)

A Weather Forecast Office (WFO) issues various public and aviation forecast and weather warnings for its area of responsibility. In support of aviation, WFOs issue terminal aviation forecasts (TAFs) and transcribed weather broadcasts (TWEBs). As MWOs, the Guam and Honolulu Hawaii WFOs issue aviation area forecasts and inflight advisories (AIRMETs, and international SIGMETs). Figures 4-1 through 4-4 show locations for which TAFs are issued. Figure 4-8 shows the TWEB routes.

Alaskan Aviation Weather Advisory Unit (AAWU)

The Alaskan Aviation Weather Unit (AAWU) is a regional aviation forecast unit located in Anchorage, Alaska. As an MWO, AAWU meteorologists prepare and issue International SIGMETs within the Alaskan FIR, as well as domestic FAs and AIRMETs for Alaska and the adjacent coastal waters. The AAWU prepares and disseminates to the FAA and the Internet a suite of graphic products, including a graphic FA and a 24- and 36-hour forecast of significant weather. The AAWU is one of nine VAACs worldwide, preparing Volcanic Ash Advisory Statements (VAAS) for the Anchorage FIR.

FEDERAL AVIATION ADMINISTRATION (FAA)

The Federal Aviation Administration (FAA) is a part of the Department of Transportation. The FAA provides a wide range of services to the aviation community. The following is a description of those FAA facilities which are involved with aviation weather and pilot services.

FLIGHT SERVICE STATIONS (FSSs)

The FAA is in the process of modernizing its Flight Service Station (FSS) program. The older, manual (or nonautomated) FSS is being consolidated into the newer, automated FSS (AFSS). With about one per state and with lines of communications radiating out from it, these new AFSSs are referred to as "hub" facilities. Pilot services provided previously by the older FSSs have been consolidated into facilities with new technology to improve pilot weather briefing services.

The FSS or AFSS provides more aviation weather briefing service than any other U.S Government service outlet. The FSS or AFSS provides preflight and inflight briefings, transcribed weather briefings, scheduled and unscheduled weather broadcasts, and furnishes weather support to flights in its area.

As a starting point for a preflight weather briefing, a pilot may wish to listen to one of the recorded weather briefings provided by an FSS or AFSS. For a more detailed briefing, pilots can contact the FSS or AFSS directly.

Transcribed Weather Broadcast (TWEB)

The transcribed weather broadcast (TWEB) provides continuous aeronautical and meteorological information on low/medium frequency (L/MF) and very high frequency (VHF) omni-directional radio range (VOR) facilities.

At TWEB equipment locations controlling two or more VORs, the one used least for ground-to-air communications, preferably the nearest VOR, may be used as a TWEB outlet simultaneously with the nondirectional radio beacon (NDB) facility.

The sequence, source, and content of transcribed broadcast material shall be:

1. Introduction.
2. Synopsis. Prepared by selected WFOs and stored in Weather Message Switching Center (WMSC).
3. Adverse Conditions. Extracted from inflight weather advisories, center weather advisories (CWAs), and alert severe weather watch bulletins (AWWs).
4. TWEB Route Forecasts. Includes the valid time of forecasts prepared by WFOs and stored in the WMSC.
5. Winds Aloft Forecast. Broadcast for the location nearest to the TWEB. The broadcast should include the levels for 3,000 to 12,000 feet, but shall always include at least two forecast levels above the surface.
6. Radar Reports. Local or pertinent radar weather reports (SDs) are used. If there is access to real-time weather radar equipment, the observed data is summarized using the SDs to determine precipitation type, intensity, movement, and height.
7. Surface Weather Reports (METARs). Surface/special weather reports are recorded, beginning with the local reports, then the remainder of the reports beginning with the first station east of true north and continuing clockwise around the TWEB location.
8. Density Altitude. Includes temperature and the "check density altitude" statement for any station with a field elevation at or above 2,000 feet MSL and meets a certain temperature criteria.
9. Pilot Weather Reports (PIREPs). PIREPs are summarized. If the weather conditions meet soliciting requirements, a request for PIREPs will be appended.
10. Alert Notices (ALNOT), if applicable.
11. Closing Statement.

Items 2, 3, 4, and 5 are forecasts and advisories prepared by the NWS and are discussed in detail in Section 4. The synopsis and route forecasts are prepared specifically for the TWEB by WFOs. Adverse conditions, outlooks, and winds/temperature aloft are adapted from inflight advisories, area forecasts, and the winds/temperature aloft forecasts. Radar reports and pilot reports are discussed in Section 3. Surface reports are discussed in Section 2.

Pilots' Automatic Telephone Weather Answering System (PATWAS)

Pilots' automatic telephone weather answering system (PATWAS) provides a continuous telephone recording of meteorological information. At PATWAS facilities where the telephone is connected to the TWEB, the information contained in the broadcast shall be in accordance with the TWEB format. PATWAS messages are recorded and updated at a minimum of every 5 hours beginning at 0600 and ending at 2200 local time using the following procedures:

1. Introduction (describing PATWAS area).
2. Adverse Conditions. Summarized inflight weather advisories, center weather advisories, alert severe weather watch bulletins, and any other available information that may adversely affect flight in the PATWAS area.
3. VFR Flight Not Recommended Statement (VNR). When current or forecast conditions, surface-based
   or aloft, would make visual flight doubtful.
4. Synopsis. Should be a reflection of current and forecast conditions using synopsis products prepared by selected WFOs or extracted from the synopsis section of the area forecast.
5. Current Conditions. Summarized current weather conditions over the broadcast area.
6. Surface Winds. Provided from local reports.
7. Forecast. Summarized forecast conditions over the PATWAS area.
8. Winds Aloft. Summarized winds aloft as forecast for the local station or as interpolated from forecasts
   of adjacent stations for levels 3,000 through 9,000 feet or a minimum of at least two
   forecast levels above the highest terrain.
9. Request for PIREPs, if applicable.
10. Alert Notices (ALNOT), if applicable.
11. Closing Announcement.

The PATWAS service holds high operational priority . This ensures the information is current and accurate. If service is reduced during the period of 2200-0600 local time, a suspension announcement is recorded including a time when the broadcast will be resumed. The Airport Facility Directory lists PATWAS telephone numbers for FSS briefing offices.

Telephone Information Briefing Service (TIBS)

Telephone information briefing service (TIBS) is provided by AFSSs and provides continuous telephone recordings of meteorological and/or aeronautical information. TIBS shall contain area and/or route briefings, airspace procedures, and special announcements, if applicable.

TIBS should also contain, but not limited to, METARs, aviation terminal forecasts (TAFs), and winds/ temperatures aloft forecasts.

Each AFSS shall provide at least four route and/or area briefings. Area briefings should encompass a 50-NM radius. Each briefing should require the pilot to access no more than two channels which shall be route- and/or area-specific. Pilots shall have access to NOTAM data through an area or route briefing on a separate channel designated specifically for NOTAMs or by access to a briefer.

TIBS service is provided 24 hours a day. Recorded information shall be updated as conditions change. Area and route forecast channels shall be updated whenever material is updated.

The order and content of the TIBS recording is as follows:

1. Introduction. Includes the preparation time and the route and/or the area of coverage.
2. Adverse Conditions. A summary of inflight weather advisories, center weather advisories, alert severe weather watch bulletins, and any other available information that may adversely affect flight in the route/area.
3. VFR Not Recommended Statement (VNR). Included when current or forecast conditions, surface or aloft, would make the flight under visual flight rules doubtful.
4. Synopsis. A brief statement describing the type, location, and movement of weather systems and/or masses which might affect the route or the area.
5. Current Conditions. A summary of current weather conditions over the route/area.
6. Density Altitude. A "check density altitude" statement will be included for any weather reporting point with a field elevation at or above 2,000 feet MSL and meets certain temperature criteria.
7. En Route Forecast. A summary of appropriate forecast data in logical order; i.e., climb out, en route, and descent.
8. Winds Aloft. A summary of winds aloft forecast for the route/area for levels through 12,000 feet.
9. Request for PIREPs, if applicable.
10. NOTAM information that affects the route/area as stated above.
11. Military Training Activity. Included in the closing announcement.
12. ALNOT Alert Announcement. If applicable.
13. Closing Announcement. Shall be appropriate for the facility equipment and the mode of operation.

Service may be reduced during the hours of 2200 and 0600 local time. During the period of reduced service, an announcement must be recorded. The Airport Facility Directory lists TIBS telephone numbers for AFSS briefing offices. A touch-tone telephone is necessary to access the TIBS program.

For those pilots already in flight and needing weather information and assistance, the following services are provided by flight service stations. They can be accessed over the proper radio frequencies printed in flight information publications.

Hazardous Inflight Weather Advisory Service (HIWAS)

The hazardous inflight weather advisory service (HIWAS) is a continuous broadcast of inflight weather advisories; i.e., SIGMETs, Convective SIGMETs, AIRMETs, AWWs, CWAs, and urgent PIREPs.

The HIWAS broadcast area is defined as the area within 150 NM of HIWAS outlets. HIWAS broadcasts shall not be interrupted/delayed except for emergency situations. The service shall be provided 24 hours a day.

An announcement shall be made if there are no hazardous weather advisories. Hazardous weather information shall be recorded if it is occurring within the HIWAS broadcast area. The broadcast shall include the following elements:

1. A statement of introduction including the appropriate area(s) and a recording time.
2. A summary of inflight weather advisories, center weather advisories, and alert severe weather watch bulletins, and any other weather not included in a current hazardous weather advisory.
3. A request for PIREPs, if applicable.
4. A recommendation to contact AFSS/FSS/FLIGHT WATCH for additional details concerning hazardous weather.

Once the HIWAS broadcast is updated, an announcement will be made once on all communications/NAVAID frequencies except emergency, and En Route Flight Advisory Service (EFAS). In the event that a HIWAS broadcast area is out of service, an announcement shall be made on all communications/NAVAID frequencies except emergency and EFAS.

En Route Flight Advisory Service (EFAS)

The en route flight advisory service (EFAS), or "Flight Watch," is a service from selected FSSs or AFSSs on a common frequency 122.0 mHz below flight level (FL) 180 and on assigned discrete frequencies to aircraft at FL180 and above. The purpose of EFAS is to provide en route aircraft with timely and pertinent weather data tailored to a specific altitude and route using the most current available sources of aviation meteorological information. Additionally, EFAS is a focal point for rapid receipt and dissemination of pilot reports. Figure 1-1 indicates the sites where EFAS and associated outlets are located. To use this service, call for flight watch. Example, "(Oakland) FLIGHT WATCH, THIS IS… "

The following paragraphs describe other FAA facilities which provide support to the aviation community.

Air Traffic Control System Command Center (ATCSCC)

The Air Traffic Control System Command Center (ATCSCC), also known as "central flow control," is located in Herndon, Virginia. ATCSCC has the mission of balancing air traffic demand with system capacity. This ensures maximum safety and efficiency for the National Airspace System while minimizing delays. The ATCSCC utilizes the Traffic Management System, aircraft situation display, monitor alert, the follow-on functions, and direct contact with the air route traffic control center (ARTCC) and terminal radar approach control facility (TRACON) traffic management units to manage flow on a national as well as local level.

Because weather is the most common reason for air traffic delays and re-routings, the ATCSCC is supported full-time by Air Traffic Control System Command Center Weather Unit Specialists (ATCSCCWUS). These specialists are responsible for the dissemination of meteorological information as it pertains to national air traffic flow management.

Air Route Traffic Control Center (ARTCC)

An air route traffic control center (ARTCC) is a facility established to provide air traffic control service to aircraft operating on IFR flight plans within controlled airspace and principally during the en route phase of flight. When equipment capabilities and controller workload permit, certain advisory/assistance services may be provided to VFR aircraft.

En route controllers become familiar with pertinent weather information and stay aware of current weather information needed to perform air traffic control duties. En route controllers shall advise pilots of hazardous weather that may impact operations within 150 NM of the controller's assigned sector or area of jurisdiction.

Center Weather Service Unit (CWSU)

The purpose of the center weather service unit (CWSU) is to provide weather consultation, forecasts, and advice to managers and staff within ARTCCs and to other supported FAA facilities. The CWSU is a joint agency aviation weather support team located at each ARTCC. The unit is composed of NWS meteorologists and FAA traffic management personnel, the latter being assigned as Weather

Coordinators. The CWSU meteorologist provides FAA traffic managers with accurate and timely weather information. This information is based on monitoring, analysis, and interpretation of real-time weather data at the ARTCC through the use of all available data sources including radar, satellite, PIREPs, and various NWS products such as TAFs and area forecasts, inflight advisories, etc. The flow or exchange of weather information between the CWSU meteorologists and the FAA personnel in the ARTCC is the responsibility of the Weather Coordinator.

Air Traffic Control Tower (ATCT)

An air traffic control tower (ATCT) is a terminal facility that uses air/ground communications, visual signaling, and other devices to provide ATC services to aircraft operating in the vicinity of an airport or on the movement area. It authorizes aircraft to land or take off at the airport controlled by the tower or to transit the Class D airspace area regardless of flight plan or weather conditions (IFR or VFR). A tower may also provide approach control services.

Terminal controllers become familiar with pertinent weather information and stay aware of current weather information needed to perform air traffic control duties. Terminal controllers shall advise pilots of hazardous weather that may impact operations within 150 NM of the controller's assigned sector or area of jurisdiction. Tower cab and approach control facilities may opt to broadcast hazardous weather information alerts only when any part of the area described is within 50 NM of the airspace under the ATCT's jurisdiction.

The responsibility for disseminating weather information is shared with the NWS at many ATCT facilities. If the responsibility is not shared, the controllers are properly certified and acting as official weather observers for the elements being reported.

An automatic terminal information service (ATIS) is a continuous broadcast of recorded information in selected terminal areas. Its purpose is to improve controller effectiveness and to relieve frequency congestion by automating the repetitive transmission of noncontrol airport/terminal area and meteorological information.

Direct User Access Terminal Service (DUATS)

The direct user access terminal system (DUATS) provides current FAA weather and flight plan filing services to U.S. Coast Guard and certified civil pilots. The computer-based system receives and stores up-to-date weather and NOTAM data from the FAA's WMSC. Pilots using a personal computer, modem, and a telephone line can access the system and request specific types of weather briefings and other pertinent data for planned flights. The pilot can also file, amend, or cancel flight plans while dialed into the system. Further information about DUATS can be obtained from any AFSS or FAA Flight Standards District Office (FSDO).

OBSERVATIONS

Weather observations are measurements and estimates of existing weather conditions both at the surface and aloft. When recorded and transmitted, an observation becomes a report; and reports are the basis of all weather analyses, forecasts, advisories, and briefings. The following paragraphs briefly describe the observation programs of the NWS and the FAA. More detailed information on each program follows.

SURFACE AVIATION WEATHER OBSERVATIONS (METARs)

Surface aviation weather observations (METARs) include weather elements pertinent to flying. A network of airport stations provides routine up-to-date surface weather information. For more information on surface aviation observation, see Section 2.

UPPER-AIR OBSERVATIONS

Upper-air weather data is received from sounding balloons (known as radiosonde observations) and pilot weather reports (PIREPs). Upper-air observations are taken twice daily at specified stations. These upper-air observations furnish temperature, humidity, pressure, and wind data, often to heights above 100,000 feet. In addition, pilots are a vital source of upper-air weather observations. In fact, aircraft in flight are the only means of directly observing turbulence, icing, and height of cloud tops. For more information on PIREPs, see Section 3. Recently some US and other international airlines have equipped their aircraft with instruments that automatically send weather observations via a satellite downlink. These are important observations which are used by NCEP in their production of forecasts.

RADAR OBSERVATIONS

The weather radar provides detailed information about precipitation, winds, and weather systems. Doppler technology allows the radar to provide measurements of winds through a large vertical depth of the atmosphere, even within "clear air." This information helps support public and aviation warning and forecast programs. Figure 7-2 shows the weather radar network across the United States.

FAA terminal doppler weather radars (TDWRs) are being installed near a number of major airports around the country. The TDWRs will be used to alert and warn airport controllers of approaching wind shear, gust fronts, and heavy precipitation which could cause hazardous conditions for landing or departing aircraft.
Also installed at major airports are the FAA airport surveillance radars. With this radar, specific locations of six different precipitation intensity levels are available for the routing of air traffic in and about a terminal location. However, the radar's primary function is for aircraft detection.

LOW-LEVEL WIND SHEAR ALERT SYSTEM (LLWAS)

The low-level wind shear alert system (LLWAS) provides pilots and controllers with information on hazardous surface wind conditions (on or near the airport) that create unsafe landing or departure conditions. LLWAS evaluates wind speed and direction from sensors on the airport periphery with center field wind data. During the time that an alert is posted, air traffic controllers provide wind shear advisories to all arriving and departing aircraft.

SATELLITE IMAGERY

Visible, infrared (IR), and other types of images (or pictures) of clouds are taken from weather satellites in orbit. These images are then made available on a near-real-time basis to NWS and FAA facilities. Satellite pictures are an important source of weather information. For more information on satellite products, see Section 3.

COMMUNICATION SYSTEM

High speed communications and automated data processing have improved the flow of weather data and products through the nation's weather network. The flow of weather information within and between agencies is becoming faster as computers and satellites are being used to facilitate the exchange of data. A new computer-based advanced weather interactive processing system (AWIPS) workstation is being deployed for the NWS. This system is replacing the current system and will allow quicker dissemination of weather information. AWIPS will be linked with the weather radars to provide better detection, observing, and forecasting of weather systems, especially severe weather.

The flow of alphanumeric weather information to the FAA service outlets is accomplished through leased lines to computer-based equipment. The NWS and FAA service outlets exchange weather information through the use of graphic products and alphanumeric information. Graphic products (weather maps) are received by FAA service outlets from NCEP through a private sector contractor. Alphanumeric information exchanged through telecommunication gateways at NWS and FAA switching centers serves to pass data between the various FAA facilities, NWS, and other users.

USERS

The ultimate users of the aviation weather service are pilots and dispatchers. Maintenance personnel may use the service to keep informed of weather that could cause possible damage to unprotected aircraft. Pilots contribute to, as well as use, the service. Pilots should send PIREPs to help fellow pilots, briefers, and forecasters. The service can be no better or more complete than the information that goes into it.

In the interest of safety and in compliance with Title 14, Code of Federal Regulations, all pilots should get a complete weather briefing before each flight. It is the responsibility of the pilot to ensure he/she has all the information needed to make a safe flight.

OBTAINING A GOOD WEATHER BRIEFING

When requesting a briefing, pilots should identify themselves as pilots and give clear and concise facts about their flight:

1. Type of flight (VFR or IFR)
2. Aircraft identification or pilot's name
3. Aircraft type
4. Departure point
5. Proposed time of departure
6. Flight altitude(s)
7. Route of flight
8. Destination
9. Estimated time en route (ETE)

With this background, the briefer can proceed directly with the briefing and concentrate on weather relevant to the flight. The weather information received depends on the type of briefing requested. A "standard" briefing should include:

1. Adverse conditions. Meteorological or aeronautical conditions reported or forecast that may influence a pilot to alter the proposed flight.
2. VFR flight not recommended (VNR). VFR flight is proposed and sky conditions or visibilities are present or forecast, surface or aloft, that, in the judgment of the AFSS/FSS briefer, would make flight under visual flight rules doubtful.
3. Synopsis. A brief statement describing the type, location, and movement of weather systems and/or air masses which might affect the proposed flight.
4. Current conditions. A summary from all available sources reporting weather conditions applicable to the flight.
5. En Route forecast. A summary from appropriate data forecast conditions applicable to the proposed flight.
6. Destination forecast. Destination forecast including significant changes expected within 1 hour before and after the ETA.
7. Winds aloft. Forecast winds aloft for the proposed route; temperature information on request.
8. NOTAMs. Provides NOTAMs pertinent to the flight.
9. ATC delays. Informs the pilot of any known ATC delays and/or flow control advisories that may affect the proposed flight.
10. Request for PIREPs. A request is made if a report of actual inflight conditions would be beneficial or when conditions meet the criteria for solicitation of PIREPs.
11. EFAS. Informs pilots of the availability of Flight Watch for weather updates.
12. Any other information the pilot requests; e.g., military training routes, etc.

An "abbreviated" briefing will be provided at the user's request:

1. To supplement mass disseminated data.
2. To update a previous briefing.
3. To request that the briefing be limited to specific information.

An "outlook" briefing will be provided when the proposed departure is 6 hours or more from the time of the briefing. Briefing will be limited to applicable forecast data needed for the proposed flight.

The FSS/AFSS's purpose is to serve the aviation community. Pilots should not hesitate to ask questions and discuss factors they do not fully understand. The briefing should be considered complete only when the pilot has a clear picture of what weather to expect. It is also advantageous for the pilot to make a final weather check immediately before departure if at all possible.

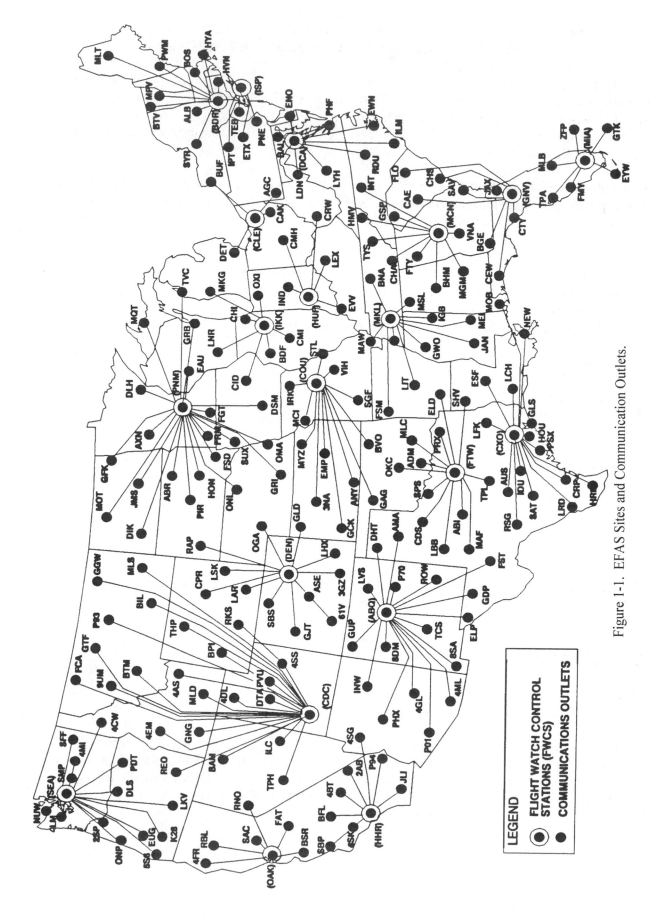

Figure 1-1. EFAS Sites and Communication Outlets.

Section 2
AVIATION ROUTINE WEATHER REPORT (METAR)

The aviation routine weather report (METAR) is the weather observer's interpretation of the weather conditions at a given site and time. The METAR is used by the aviation community and the National Weather Service (NWS) to determine the flying category - visual fight rules (VFR), marginal VFR (MVFR), or instrument flight rules (IFR) - of the airport, as well as produce the Aviation Terminal Forecast (TAF). (See Section 4.)

Although the METAR code is being adopted worldwide, each country is allowed to make modifications or exceptions to the code for use in that particular country. The U.S.A. reports temperature and dew point in degrees Celsius and continues to use current units of measurement for the remainder of the report.

The elements in the body of a METAR report are separated with a space. The only exception is temperature and dew point that are separated with a solidus (/). When an element does <u>not</u> occur, or cannot be observed, the preceding space and that element are omitted from that particular report. A METAR report contains the following elements in order as presented:

1. Type of report
2. ICAO station identifier
3. Date and time of report
4. Modifier (as required)
5. Wind
6. Visibility
7. Runway visual range (RVR) (as required)
8. Weather phenomena
9. Sky condition
10. Temperature/dew point group
11. Altimeter
12. Remarks (RMK) (as required)

The following paragraphs describe the elements in a METAR report. A sample report will accompany each element with the subject element highlighted.

TYPE OF REPORT

METAR KLAX 140651Z AUTO 00000KT 1SM R35L/4500V6000FT -RA BR BKN030 10/10 A2990 RMK AO2

There are two types of reports: The METAR and the aviation selected special weather report (SPECI). The METAR is observed hourly between 45 minutes after the hour till the hour and transmitted between 50 minutes after the hour till the hour. It will be encoded as a METAR even if it meets SPECI criteria. The SPECI is a non-routine aviation weather report taken when any of the SPECI criteria have been observed. The criteria are listed in Table 2-1, "SPECI Criteria."

Table 2-1 SPECI Criteria

Report Element	Criteria
Wind	Wind direction changes by 45 degrees or more in less than 15 minutes and the wind speed is 10 knots or more throughout the windshift.
Visibility	Surface visibility as reported in the body of the report decreases to less than, or if below, increases to equal or exceeds: 3,2, or 1mile or the lowest standard instrument approach procedure minimum as published in the National Ocean Service U.S Instrument Procedures. If none is published use ½ mile.
RVR	Changes to above or below 2,400 feet
Tornado, Funnel Cloud, Waterspout	When observed or when disappears from sight (ends)
Thunderstorm	Begins or ends
Precipitation	When freezing precipitation or ice pellets begin, end, or change intensity or hail begins or ends
Squalls	When they occur
Ceilings	The ceiling forms or dissipates below, decreases to less than, or if below, increases to equal or exceeds: 3,000, 1,500, 1,000, or 500 feet or the lowest standard instrument approach procedure minimum as published in the National Ocean Service U.S Instrument Procedures. If none is published use 200 feet.
Sky Condition	A layer of clouds or obscuring phenomenon aloft that forms below 1,000 feet
Volcanic Eruption	When an eruption is first noted
Aircraft Mishap	Upon notification of an aircraft mishap, unless there has been an intervening observation
Miscellaneous	Any other meteorological situation designated by the agency, or which, in the opinion of the observer, is critical

ICAO STATION IDENTIFIER

METAR **KLAX** 140651Z AUTO 00000KT 1SM R35L/4500V6000FT -RA BR BKN030 10/10 A2990 RMK AO2

The METAR code uses International Civil Aviation Organization (ICAO) four-letter station identifiers that follow the type of report. In the conterminous United States, the three-letter identifier is prefixed with K. For example SEA (Seattle) becomes KSEA. Elsewhere, the first one or two letters of the ICAO identifier indicate in which region of the world and country (or state) the station is located. Pacific locations such as Alaska, Hawaii, and the Mariana Islands start with P followed by an A, H, or G respectively. The last two letters reflect the specific reporting station identification. If the location's three-letter identification begins with an A, H, or G, the P is added to the beginning. If the location's three-letter identification does not begin with an A, H, or G, the last letter is dropped and the P is added to the beginning.

Examples:
ANC (Anchorage, AK) becomes PANC.
OME (Nome, AK) becomes PAOM.
HNL (Honolulu, HI) becomes PHNL.
KOA (Keahole Point, HI) becomes PHKO.
UAM (Anderson AFB, Guam) becomes PGUA.

Canadian station identifiers start with C.

Example:
Toronto, Canada, is CYYZ.

Mexican and western Caribbean station identifiers start with M.

Examples:
Mexico City, Mexico, is MMMX.
Guantanamo, Cuba, is MUGT.
Santo Domingo, Dominican Republic, is MDGM.
Nassau, Bahamas, is MYNN.

The identifier for the eastern Caribbean is T, followed by the individual country's letter.

Example:
San Juan, Puerto Rico, is TJSJ.

For a complete worldwide listing, see ICAO Document 7910, "Location Indicators."

DATE AND TIME OF REPORT

METAR KLAX **140651Z** AUTO 00000KT 1SM R35L/4500V6000FT -RA BR BKN030 10/10 A2990 RMK AO2

The date and time the observation is taken are transmitted as a six-digit date/time group appended with **Z** to denote Coordinated Universal Time (UTC). The first two digits are the date followed with two digits for hour and two digits for minutes. If a report is a correction to a previously disseminated erroneous report, the time entered on the corrected report shall be the same time used in the report being corrected.

MODIFIER (AS REQUIRED)

METAR KLAX 140651Z **AUTO** 00000KT 1SM R35L/4500V6000FT -RA BR BKN030 10/10 A2990 RMK AO2

The modifier element, if used, follows the date/time element. The modifier, **AUTO**, identifies a METAR/SPECI report as an automated weather report with no human intervention. If AUTO is shown in the body of the report, AO1 or AO2 will be encoded in the remarks section of the report to indicate the type of precipitation sensor used at the station. A remark of AO1 indicates a report from a station that does <u>not</u> have a precipitation discriminator, and an AO2 remark indicates a report from a station which has a precipitation discriminator. The absence of AUTO indicates that the report was made manually or the automated report had human augmentation/backup.

The modifier, **COR**, identifies a corrected report that is sent out to replace an earlier report with an error.

Example:
METAR KLAX 140651Z **COR**...

WIND

METAR KLAX 140651Z AUTO **00000KT** 1SM R35L/4500V6000FT -RA BR BKN030 10/10 A2990 RMK AO2

The wind element is reported as a five-digit group (six digits if speed is over 99 knots). The first three digits are the direction from which the wind is blowing in tens of degrees referenced to true north. Directions less than 100 degrees are preceded with a zero. The next two digits are the average speed in knots, measured or estimated, or if over 99 knots, the next three digits.

Example:
340105KT

If the wind speed is less than 3 knots, the wind is reported as calm, 00000KT. If the wind is gusty, 10 knots or more between peaks and lulls, **G** denoting gust is reported after the speed followed by the highest gust reported. The abbreviation **KT** is appended to denote the use of knots for wind speed. Other countries may use kilometers per hour or meters per second.

If the wind direction is variable by 60 degrees or more and the speed is greater than 6 knots, a variable group consisting of the extremes of the wind directions separated by **V** will follow the wind group.

Example:
08012G25KT 040V120

The wind direction may also be considered variable if the wind speed is 6 knots or less and is varying in direction (60-degree rule does not apply). This is indicated with the contraction **VRB**.

Example:
VRB04KT

WIND REMARKS

At facilities that have a wind recorder or at automated weather reporting systems, whenever the peak wind exceeds 25 knots, **PK WND** will be included in the Remarks element in the next report. The peak wind remark includes three digits for direction and two or three digits for speed followed by the time in hours and minutes of occurrence. If the hour can be inferred from the report time, only the minutes are reported.

Example:
PK WND 28045/15

When a windshift occurs, **WSHFT** will be included in the Remarks element followed by the time the windshift began (with only minutes reported, if the hour can be inferred from the time of observation). A windshift is indicated by a change in wind direction of 45 degrees or more in less than 15 minutes with sustained winds of 10 knots or more throughout the windshift. The contraction, **FROPA**, may be entered following the time if the windshift is the result of a frontal passage.

Example:
WSHFT 30 FROPA

VISIBILITY

METAR KLAX 140651Z AUTO 00000KT **1SM** R35L/4500V6000FT -RA BR BKN030 10/10 A2990 RMK AO2

Prevailing visibility is reported in statute miles followed by a space, fractions of statute miles, as needed, and the letters **SM**. Other countries may use meters or kilometers. Prevailing visibility is considered representative of the visibility conditions at the observing site. Prevailing visibility is the greatest visibility equaled or exceeded throughout at least half the horizon circle, which need not be continuous. When visibilities are less than 7 miles, the restriction to visibility will be shown in the weather element. The only exception to this rule is that if volcanic ash, low drifting dust, sand, or snow is observed, it is reported, even if it does not restrict visibility to less than 7 miles.

VISIBILITY REMARKS

If tower or surface visibility is less than 4 statute miles, the lesser of the two will be reported in the body of the report; the greater will be reported in the Remarks element.

Example:
TWR VIS 1 1/2 or SFC VIS 1 1/2

Automated reporting stations will show visibility less than 1/4 statute mile as **M1/4SM** and visibility 10 or greater than 10 statute miles as **10SM**.

For automated reporting stations having more than one visibility sensor, site-specific visibility (which is lower than the visibility shown in the body) will be shown in the Remarks element.

Example:
VIS 2 1/2 RWY 11

When the prevailing visibility rapidly increases or decreases by 1/2 statute mile or more during the observation, and the average prevailing visibility is less than 3 statute miles, the visibility is variable. Variable visibility is shown in the Remarks element with the minimum and maximum visibility values separated by a V.

Example:
VIS 1/2V2

Sector visibility is shown in the Remarks element when it differs from the prevailing visibility and either the prevailing or sector visibility is less than 3 miles.

Example:
VIS NE 2 1/2

RUNWAY VISUAL RANGE (RVR) (AS REQUIRED)

METAR KLAX 140651Z AUTO 00000KT 1SM **R35L/4500V6000FT** -RA BR BKN030 10/10 A2990 RMK AO2

Runway visual range (RVR) follows the visibility element. RVR, when reported, is in the following format: **R** identifies the group; followed by the runway heading and, if needed, the parallel runway designator; next is a solidus (**/**); last is the visual range in feet (meters in other countries) indicated by "**FT**." RVR is shown in a METAR/SPECI if the airport has RVR equipment and whenever the prevailing visibility is 1 statute mile or less and/or the RVR value is 6,000 feet or less. When the RVR varies by more than one reportable value, the lowest and highest values are shown with **V** between them.

Example:
R35L/4500V6000FT

When the observed RVR is above the maximum value which can be determined by the system, it should be reported as P6000 where 6,000 is the maximum value for this system. When the observed RVR is below the minimum value which can be determined by the system, it should be reported as M0600 where 600 is the minimum value for this system.

Example:
R27/P6000FT and R12C/M0600FT

WEATHER PHENOMENA

METAR KLAX 140651Z AUTO 00000KT 1SM R35L/4500V6000FT **-RA BR** BKN030 10/10 A2990 RMK AO2

Weather phenomena is broken into two categories: qualifiers and weather phenomena.

QUALIFIERS

Intensity

Intensity may be shown with most precipitation types.

A "-"denotes a light intensity level, no symbol denotes a moderate intensity level, and a "+" denotes a heavy intensity level. When more than one type of precipitation is present, the intensity refers to the first precipitation type (most predominant). (See Table 2-2.)

Example:
+SHRASN indicates heavy rainshowers and snow.

Table 2-2 Intensity Qualifiers

Intensity Qualifiers	
-	Light
	Moderate
+	Heavy

Proximity

Proximity is applied to and reported only for weather phenomena occurring in the vicinity of the airport. Vicinity of the airport is defined to be between 5 and 10 miles of the usual point of observation for obscuration and just beyond to point of observation to 10 miles for precipitation. It is denoted by **VC**. Intensity and VC will never be shown in the same group.

Examples:
VCSH indicates showers in the vicinity of the airport.
VCFG indicates fog in the vicinity of the airport.

Descriptor

The eight descriptors shown in Table 2-3 further identify weather phenomena and are used with certain types of precipitation and obscurations. Although **TS** and **SH** are used with precipitation and may be preceded with an intensity symbol, the intensity still applies to the precipitation and not the descriptor.

Example:
+TSRA is a thunderstorm with heavy rain and not a heavy thunderstorm with rain.

Table 2-3 Descriptor Qualifiers

Descriptor	
MI[1]	Shallow
BC[2]	Patches
DR[3]	Low drifting
BL[4]	Blowing
SH	Showers
TS	Thunderstorm
FZ	Freezing
PR	Partial

[1]MI shall be used only to further describe fog that has little vertical extent (less than 6 feet).
[2]BC shall be used only to further describe fog that has little vertical extent and reduces horizontal visibility.
[3]DR shall be used when dust, sand, or snow is raised by the wind to less than 6 feet.
[4]BL shall be used when dust, sand, snow, and/or spray is raised by the wind to a height of 6 feet or more.

WEATHER PHENOMENA

If more than one significant weather phenomenon is observed, entries will be listed in the following order: Tornadic activity, thunderstorms, and weather phenomena in order of decreasing predominance (i.e., the most dominant type is reported first).

If more than one significant weather phenomenon is observed, except precipitation, separate weather groups will be shown in the report. No more than three weather groups will be used to report weather phenomena at or in the vicinity of the station. If more than one type of precipitation is observed, the appropriate contractions are combined into a single group with the predominant type being reported first. In such a group, any intensity will refer to the first type of precipitation in the group.
Refer to Table 2-4 while reading the rest of this section.

Examples:
TSRA indicates thunderstorm with moderate rain.
+SHRA indicates heavy rainshowers.
-FZRA indicates light freezing rain.

Precipitation

The types of precipitation in the METAR code are shown in Table 2-4. Precipitation is any form of water particle, whether liquid or solid, that falls from the atmosphere and reaches the ground.

Examples:

GR is used to indicate hail ¼ inch in diameter or larger.

GS is used to indicate hail less than ¼ inch in diameter.

UP is unknown precipitation and is used only at automated sites. This occurs when light precipitation is falling but the precipitation discriminator cannot determine the precipitation type. This situation usually occurs when rain and snow are falling at the same time.

Obscurations

The types of obscuration phenomena in the METAR code are shown in Table 2-4. They are any phenomena in the atmosphere, other than precipitation, that reduce horizontal visibility.

Examples:

BR is used to indicate mist restricting visibility and is used only when the visibility is from 5/8 mile to 6 miles.

FG is used to indicate fog restricting visibility and is used only when visibility is less than 5/8 mile.

Other

The other weather phenomena, shown in the table, are reported when they occur.

Examples:

SQ is a sudden increase in wind speed of at least 16 knots, the speed rising to 22 knots or more, and lasting at least 1 minute.

+FC is used to denote a tornado or waterspout.

FC is used to denote a funnel cloud.

Table 2-4 Weather Phenomena

Precipitation	Obscuration	Other
DZ Drizzle	**BR Mist**	**PO Dust/Sand whirls**
RA Rain	**FG Fog**	**SQ Squalls**
SN Snow	**DU Dust**	**FC Funnel cloud**
SG Snow grains	**SA Sand**	**+FC Tornado or Waterspout**
IC Ice crystals	**HZ Haze**	**SS Sandstorm**
PL Ice pellets	**PY Spray**	**DS Dust storm**
GR Hail	**VA Volcanic ash**	
GS Small hail or Snow pellets	**FU Smoke**	
UP Unknown precipitation		

Weather Begins/Ends

When weather begins or ends, the Remarks element will show the beginning and ending times of any type of precipitation or thunderstorms. Types of precipitation may be combined if beginning or ending times are the same.

Example:
RAB05E30SNB30E45 This means that rain began at 5 minutes past the hour and ended at 30 minutes past the hour, snow began at 30 minutes past the hour and ended at 45 minutes past the hour.

Example:
TSB05E45 This means a thunderstorm began at 5 minutes past the hour and ended at 45 minutes past the hour.

Hailstone Size

When hailstones, **GR**, are shown in the body of a report, the largest hailstone size is shown in the Remarks element in 1/4-inch increments and identified with the contraction GR. Hailstones less than 1/4 inch are shown in the body of a report as **GS** and no remarks are entered indicating hailstone size.

Example:
GR 1 ¾

SKY CONDITION

METAR KLAX 140651Z AUTO 00000KT 1SM R35L/4500V6000FT -RA BR **BKN030** 10/10 A2990 RMK AO2

Sky condition is reported in the following format:

Amount/Height/Type (as required) or Indefinite Ceiling/Height (Vertical Visibility)

AMOUNT

A clear sky, a layer of clouds, or an obscuring phenomenon is reported by one of the six sky cover contractions. (See Table 2-5.) When more than one layer is reported, they are reported in ascending order of height. For each layer above a lower layer or layers, the sky cover for that higher layer will be the total sky cover that includes that higher layer and all lower layers. In other words, the summation of the cloud layers from below and at that higher level determines what sky cover is reported. This summation includes both clouds and obscuration. The amount of sky cover is reported in eighths of the sky, using the contractions in Table 2-5.

Table 2-5 Reportable Contractions for Sky Cover

Reportable Contractions	Meaning	Summation Amount
*SKC or CLR	Clear	0 or 0 below 12,000
FEW	Few	>0 but $\leq$ 2/8
SCT	Scattered	3/8-4/8
BKN	Broken	5/8-7/8
OVC	Overcast	8/8
VV	Vertical Visibility (indefinite ceiling)	8/8

SKC will be reported at manual stations. The abbreviation **CLR** shall be used at automated stations when no clouds below 12,000 feet are detected.

Note: For aviation purposes, the ceiling is defined as the height (AGL) of the lowest broken or overcast layer aloft or vertical visibility into an obscuration.

HEIGHT

Cloud bases are reported with three digits in hundreds of feet above ground level.

Example:
SCT020

Clouds above 12,000 feet cannot be detected by automated reporting systems. At reporting stations located in the mountains, if the cloud layer is below the station level, the height of the layer will be shown as three solidi (///).

Example:
SCT///

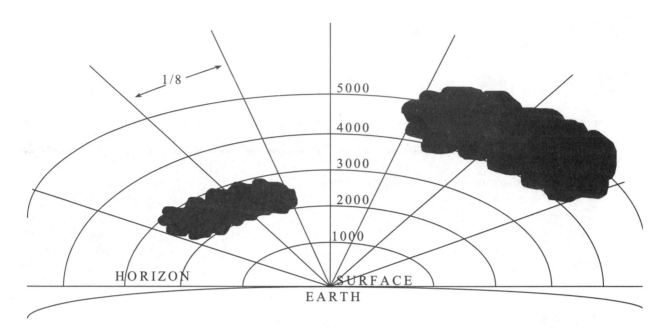

Figure 2-1. Few sky cover at 2,000 feet (2/8) and scattered sky cover at 4,000 feet (4/8). The summation of sky cover would be coded as FEW020 SCT040.

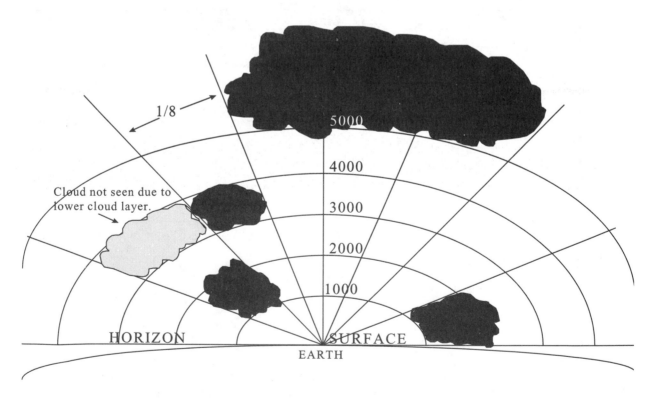

Figure 2-2. The sky cover consists of few clouds at 1,000 feet (2/8), scattered clouds at 3,000 feet (3/8), and broken clouds at 5,000 feet (6/8). This is coded as FEW010 SCT030 BKN050.

TYPE (AS REQUIRED)

If towering cumulus clouds, **TCU**, or cumulonimbus clouds, **CB**, are present, they are reported after the height that represents their base.

Example:
 BKN025CB or SCT040TCU

Figure 2-3. Towering Cumulus (TCU). The significance of this cloud is that it indicates the atmosphere in the lower altitudes is unstable and conducive to turbulence. (Photo courtesy of National Severe Storms Laboratory/University of Oklahoma.)

Figure 2-4. Cumulonimbus (CB). The anvil portion of a CB is composed of ice crystals. The CB or thunderstorm cloud contains most types of aviation weather hazards, particularly turbulence, icing, hail, and low-level wind shear (LLWS). (Photo courtesy of Doug Streu.)

INDEFINITE CEILING/HEIGHTS (VERTICAL VISIBILITY)

The height into an indefinite ceiling is preceded with VV followed by three digits indicating the vertical visibility in hundreds of feet above ground level. The layer is spoken as "indefinite ceiling" and indicates total obscuration.

Example:
VV002

Partial Obscurations

The amount of obscuration is reported in the body of the METAR when the sky is partially obscured by a surface-based phenomenon by indicating the amount of obscuration as **FEW**, **SCT**, or **BKN** followed with three zeros **(000)**. The type of obscuring phenomenon is stated in the Remarks element and precedes the amount of obscuration and three zeros. For example, if fog is hiding >1/8 to 2/8 of the sky, it will be coded in the body of the METAR as "FEW000." Because the fog is partially obscuring the sky, a remark is required. (See Figure 2-5.)

Example of Remark:
FG FEW000.

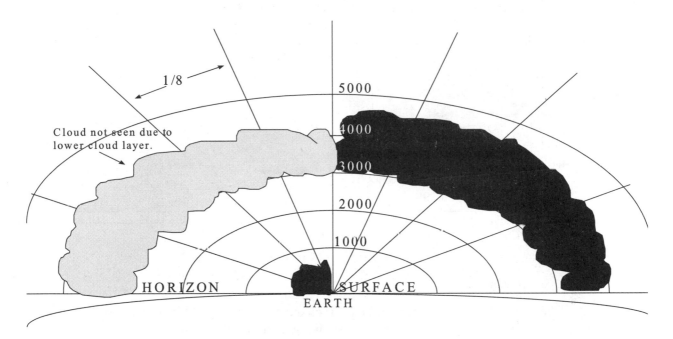

Figure 2-5. The sky cover consists of surface-based obscuration and an overcast layer at 3,000 feet. This is coded as SCT000 OVC030 with FG SCT000 in remarks.

The sky cover and ceiling, as determined from the ground, represent as nearly as possible what the pilot should experience in flight. In other words, a pilot flying at or above a reported ceiling layer (BKN-OVC) should see less than half the surface below. A pilot descending through a surface-based total obscuration should first see the ground directly below from the height reported as vertical visibility into the obscuration. However, due to the differing viewing points of the pilot and the observer, the observed values and what the pilot sees do not always exactly agree. Figure 2-6 illustrates the effect of an obscured sky on the vision from a descending aircraft.

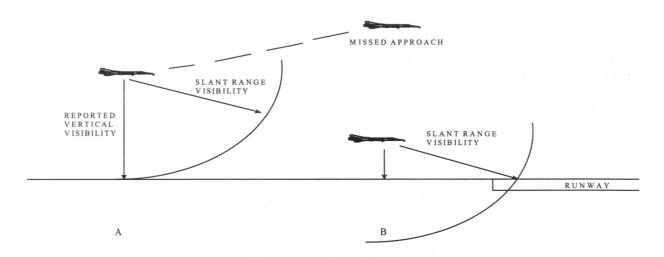

Figure 2-6. The obscuration limits runway acquisition due to slant range effects (A). This pilot would be able to see the ground but not the runway. The pilot will not be able to see the runway until position B, which is at a much lower altitude. If position A represented approach minimums, the approach could not be continued and a missed approach must be executed.

ADDITIONAL SKY CONDITION REMARKS

Whenever the ceiling is below 3,000 feet and is variable, the remark **CIG** will be shown in the Remarks element followed with the lowest and highest ceiling heights separated with a **V**.

Example:
CIG 005V010

When an automated station uses meteorological discontinuity sensors, site-specific sky conditions that differ from the ceiling height in the body of the report will be shown in the Remarks element.

Example:
CIG 002 RWY 11

When a layer is varying in sky cover, the variability range will be shown in the Remarks element. If there is more than one cloud layer of the same coverage, the variable layer will be identified by including the layer height.

Example:
BKN014 V OVC

When significant clouds are observed, they are shown in the Remarks element. The following cloud types are shown:

Towering cumulus, **TCU**, and direction from the station.

Example:
TCU W

Cumulonimbus, **CB**; or cumulonimbus mammatus, **CBMAM;** direction from the station; and direction of movement (if known). If the clouds are beyond 10 miles from the airport, **DSNT** will indicate that they are distant. (See Figure 2-7.)

Examples:
CB DSNT E or CBMAM E MOV NE
(For TCU and CB see Figures 2-3 and 2-4.)

Figure 2-7. Cumulonimbus Mammatus (CBMAM). This characteristic cloud can result from violent up- and downdrafts. This cloud type indicates possible severe or greater turbulence. (Photo courtesy of Grant Goodge taken at Asheville, NC on 4/15/87.)

Altocumulus castellanus, **ACC**; standing lenticular (stratocumulus, **SCSL**; altocumulus, **ACSL**; or cirrocumulus, **CCSL**); or rotor clouds, **ROTOR CLD,** will show a remark describing the clouds (if needed) and the direction from the station.

Examples:
ACC NW or ACSL SW
(Figure 2-8 for ACC; see Figure 2-9 for standing lenticular clouds.)

Figure 2-8. Altocumulus Castellanus (ACC). ACC indicates unstable conditions aloft, but not necessarily below the base of the cloud. (Photo courtesy of National Severe Storms Laboratory/University of Oklahoma.)

Figure 2-9. Standing Lenticular Altocumulus (ACSL). These clouds are characteristic of the standing or mountain wave. Similar clouds are rotor clouds and standing lenticular cirrocumulus (CCSL). The rotor clouds are usually at a lower altitude than the ACSL. CCSL are whiter and at a higher altitude. All three cloud types are indicative of possible severe or greater turbulence. (Photo courtesy of Grant Goodge taken at Concord, CA in 1970.)

TEMPERATURE/DEW POINT GROUP

METAR KLAX 140651Z AUTO 00000KT 1SM R35L/4500V6000FT -RA BR BKN030 **10/10** A2990 RMK AO2

Temperature/dew point are reported in a two-digit form in whole degrees Celsius separated by a solidus (/). Temperatures below zero are prefixed with **M**. If the temperature is available but the dew point is missing, the temperature is shown followed by a solidus. If the temperature is missing, the group is omitted from the report.

ALTIMETER

METAR KLAX 140651Z AUTO 00000KT 1SM R35L/4500V6000FT -RA BR BKN030 10/10 **A2990** RMK AO2

The altimeter element follows the temperature/dew point group and is the last element in the body of a METAR or SPECI report. It is reported in a four-digit format representing tens, units, tenths, and hundredths of inches of mercury prefixed with **A**. The decimal point is <u>not</u> reported or stated.

ALTIMETER REMARKS

When the pressure is rising or falling rapidly at the time of observation, Remarks element will show **PRESRR** or **PRESFR** respectively.

Some stations also include the sea-level pressure (which is different from altimeter). It is shown in the Remarks element as **SLP** being the remark identifier followed by the sea-level pressure in hectopascals (h/Pa), a unit of measurement equivalent to millibar (mb).

Example:
SLP982

REMARKS (RMK) (AS REQUIRED)

METAR KLAX 140651Z AUTO 00000KT 1SM R35L/4500V6000FT -RA BR BKN030 10/10 A2990
RMK AO2

Remarks will be included in all observations, when appropriate, in the order as presented in Table 2-6. The contraction **RMK** follows the altimeter in the body and precedes the actual remarks. Time entries will be shown as minutes past the hour if the time reported occurs during the same hour the observation is taken. If the hour is different, hours and minutes will be shown. Location of phenomena within 5 statute miles of the point of observation will be reported as at the station. Phenomena between 5 and 10 statute miles will be reported as in the vicinity, **VC**. Phenomena beyond 10 statute miles will be shown as distant, **DSNT**. Direction of phenomena will be indicated with the eight points of the compass (i.e., N, NE, E, SE, S, SW, W, NW). Distance remarks are in statute miles except for automated lightning remarks that are in nautical miles. Movement of clouds or weather will be indicated by the direction toward which the phenomenon is moving.

There are two categories of remarks: automated, manual, and plain language; and additive and automated maintenance data.

AUTOMATED, MANUAL, AND PLAIN LANGUAGE REMARKS CATEGORY

This group of remarks may be generated from either manual or automated weather reporting stations and generally elaborate on parameters reported in the body of the report. (See Table 2-6.)

Table 2-6 Automated, Manual, and Plain Language Remarks

Remarks	Examples of Remarks
1. Volcanic Eruptions	MT. AUGUSTINE VOLCANO 70 MILES SW ERUPTED 231505 LARGE ASH CLOUD EXTENDING TO APRX 30000 FEET MOVING NE
2. Tornado, Funnel Cloud, or Waterspout	TORNADO B13 6 NE
3. Automated Station Type	AO1 or AO2
4. Peak Wind	PK WND 28045/15
5. Windshift	WSHFT 30 FROPA
6. Tower Visibility or Surface Visibility	TWR VIS 1 ½ or SFC VIS 1 ½
7. Variable Prevailing Visibility	VIS 1/2V2
8. Sector Visibility	VIS NE 2 ½
9. Visibility at Second Site	VIS 2 ½ RWY 11
10. Lightning	OCNL LTGICCG OHD or FRQ LTGICCCCG W
11. Beginning and Ending of Precipitation	RAB05E30SNB20E55
12. Beginning and Ending of Thunderstorm	TSB05E30
13. Thunderstorm Locations	TS SE MOV NE
14. Hailstone Size	GR 1 ¾
15. Virga	VIRGA NE (See Figure 2-10.)
16. Variable Ceiling Height	CIG 005V010
17. Obscurations	FU BKN000
18. Variable Sky Condition	BKN014 V OVC
19. Significant Cloud Types	CB W MOV E or CBMAM S MOV E or TCU W or ACC NW or ACSL SW-W
20. Ceiling Height at Second Location	CIG 002 RWY 11
21. Pressure Rising or Falling Rapidly	PRESRR or PRESFR
22. Sea-Level Pressure	SLP982
23. Aircraft Mishap	(ACFT MSHP)
24. No SPECI Report Taken	NOSPECI
25. Snow Increasing Rapidly	SNINCR 2/10
26. Other Significant Information	Any other information that will impact aviation operations

FIGURE 2-10. Virga. Virga is precipitation falling from a cloud but evaporating before reaching the ground. Virga results when air below the cloud is very dry and is common in the western part of the country. Virga associated with showers suggests strong downdrafts with possible moderate or greater turbulence. (Photo courtesy of Grant Goodge.)

ADDITIVE AND AUTOMATED MAINTENANCE DATA REMARKS CATEGORY

Additive data groups are reported only at designated stations. The maintenance data groups are reported only from automated weather reporting stations. The additive data and maintenance groups are <u>not</u> used by the aviation community

EXAMPLES OF METAR REPORTS AND EXPLANATIONS:

METAR KMKL 021250Z 33018KT 290V360 1/2SM R31/2600FT SN BLSN FG VV008 00/M03 A2991 RMK RAESNB42 SLPNO T00111032

METAR	aviation routine weather report
KMKL	Jackson, TN
021250Z	date 02, time 1250 UTC
33018KT	wind 330 at 18 knots
290V360	wind direction variable between 290 and 360 degrees
1/2SM	visibility one-half statute mile
R31/2600FT	runway 31, RVR 2600
SN	moderate snow
BLSN FG	blowing snow and fog
VV008	indefinite ceiling 800
00/M03	temperature 0°C, dew point -3°C
A2991	altimeter 2991
RMK	remarks
RAESNB42	rain ended at four two, snow began at four two
SLPNO	sea-level pressure not available
T00111032	temperature 1.1°C, dew point -3.2°C

The following is an example of the phraseology used to relay this report to a pilot. Optional phrases or words are shown in parentheses.

"Jackson (Tennessee), (one two five zero observation), wind three three zero at one eight, wind variable between two niner zero and three six zero, visibility one-half, runway three one RVR, two thousand six hundred, heavy snow, blowing snow, fog, indefinite ceiling eight hundred, temperature zero, dew point minus three, altimeter two niner niner one."

METAR KSFO 031453Z VRB02KT 7SM MIFG SKC 15/14 A3012 RMK SLP993 6//// T01500139 56012

METAR	aviation routine weather report
KSFO	San Francisco, CA
031453Z	date 03, time 1453 UTC
VRB02KT	wind variable at 2 knots
7SM	visibility 7 statute miles
MIFG	shallow fog
SKC	clear
15/14	temperature 15°C, dew point 14°C
A3012	altimeter 3012
RMK	remarks
SLP993	sea-level pressure 999.3 hectopascals
6////	an indeterminable amount of precipitation has occurred over the last 3 hours
T01500139	temperature 15.0°C, dew point 13.9°C
56012	atmospheric pressure lower since previous 3 hours ago

The following is an example of the phraseology used to relay this report to a pilot. Optional phrases or words are shown in parentheses.

"San Francisco (one four five three observation), wind variable at two, visibility seven, shallow fog, clear, temperature one five, dew point one four, altimeter three zero one two."

SPECI KCVG 312228Z 28024G36KT 3/4SM +TSRA SQ BKN008 OVC020CB 28/23 A3000 RMK TSB24 TS OHD MOV E

SPECI	aviation selected special weather report
KCVG	Covington, KY
312228Z	date 31, time 2228 UTC
28024G36KT	wind 280 at 24, gusts 36 knots
3/4SM	visibility three-quarters statute mile
+TSRA SQ	thunderstorm with heavy rain and squalls
BKN008 OVC020CB	ceiling 800 broken, 2,000 overcast, cumulonimbus
28/23	temperature 28°C, dew point 23°C
A3000	altimeter 3000
RMK	remarks
TSB24	thunderstorm began at two four
TS OHD MOV E	thunderstorm overhead moving east

The following is an example of the phraseology used to relay this report to a pilot. Optional phrases or words are shown in parentheses.

"Covington (Kentucky), special report, two eight observation, wind two eight zero at two four, gusts three six, visibility three-quarters, thunderstorm, heavy rain, squall, ceiling eight hundred broken, two thousand overcast, cumulonimbus, temperature two eight, dew point two three, altimeter three zero zero zero, thunderstorm began two four, thunderstorm overhead, moving east."

More examples without phraseology:

METAR KLAX 140651Z AUTO 00000KT 10SM -RA SCT080 12/05 A2990 RMK AO2

METAR	aviation routine weather report
KLAX	Los Angeles, CA
140651Z	date 14, time 0651 UTC
AUTO	automated site
00000KT	calm winds
10SM	visibility 10 statute miles
-RA	light rain
SCT080	8,000 scattered
12/05	temperature 12°C, dew point 5°C
A2990	altimeter 2990
RMK	remarks
AO2	automated observation with precipitation discriminator

SPECI KDEN 241310Z 09014G35KT 1/4SM +SN FG VV002 01/01 A2975 RMK AO2 TWR VIS ½ RAE08SNB08

SPECI	aviation selected special weather report
KDEN	Denver, CO
241310Z	date 24, time 1310 UTC
09014G35KT	wind 090 at 14, gusts to 35 knots
1/4SM	visibility one-quarter statute mile
+SN FG	heavy snow and fog
VV002	indefinite ceiling 200
01/01	temperature 1°C, dew point 1°C
A2975	altimeter 2975
RMK	remarks
AO2	automated observation with precipitation discriminator
TWR VIS 1/2	tower visibility one-half
RAE08SNB08	rain ended and snow began at 08 minutes after the hour

METAR KSPS 301656Z 06014KT 020V090 3SM -TSRA FEW040 BKN060CB 12/ A2982 RMK OCNL LTGICCG NE TSB17 TS E MOV NE PRESRR SLP093

METAR	aviation routine weather report
KSPS	Wichita Falls, TX
301656Z	date 30, time 1656 UTC
06014KT 020V090	wind 060 at 14 knots varying between 020 and 090 degrees
3SM	visibility 3 statute miles
-TSRA	thunderstorm with light rain
FEW040 BKN060CB	few clouds at 4,000, ceiling 6,000 broken, cumulonimbus
12/	temperature 12°C (dew point is missing)
A2982	altimeter 2982
RMK	remarks
OCNL LTGICCG NE	occasional lightning in cloud, cloud-to-ground northeast
TSB17	thunderstorm began 17
TS E MOV NE	thunderstorm east moving northeast
PRESRR	pressure rising rapidly
SLP093	sea-level pressure 1009.3 hectopascals

SPECI KBOS 051237Z VRB02KT 3/4SM R15R/4000FT BR OVC004 05/05 A2998 RMK AO2 CIG 002V006

SPECI	aviation selected special weather report
KBOS	Boston, MA
051237Z	date 5, time 1237 UTC
VRB02KT	variable wind at 2 knots
3/4SM	visibility three-quarters statute mile
R15R/4000FT	runway visual range on runway 15R 4,000 feet
BR	mist
OVC004	ceiling 400 overcast
05/05	temperature 5°C, dew point 5°C
A2998	altimeter 2998
RMK	remarks
AO2	automated observation with precipitation discriminator
CIG 002V006	ceiling variable 200 to 600

Section 3
PILOT AND RADAR REPORTS, SATELLITE PICTURES, AND RADIOSONDE ADDITIONAL DATA (RADATs)

The preceding section explained the decoding of METAR reports. However, these "spot" reports are only one facet of the total current weather picture. Pilot and radar reports, satellite pictures, and radiosonde additional data (RADATs) help to fill the gaps between stations.

PILOT WEATHER REPORTS (PIREPs)

No observation is more timely than the one made from the flight deck. In fact, aircraft in flight are the only means of observing icing and turbulence. Other pilots welcome pilot weather reports (PIREPs) as well as do the briefers and forecasters. A PIREP always helps someone and becomes part of aviation weather. Pilots should report any observation that may be of concern to other pilots. Also, if conditions were forecasted but were not encountered, a pilot should also provide a PIREP. This will help the NWS to verify forecast products and create accurate products for the aviation community. Pilots should help themselves, the aviation public, and the aviation weather forecasters by providing PIREPs.

A PIREP is transmitted in a prescribed format (see Table 3-1). Required elements for all PIREPs are type of report, location, time, flight level, aircraft type, and at least one weather element encountered. When not required, elements without reported data are omitted. All altitude references are mean sea level (MSL) unless otherwise noted. Distance for visibility is in statute miles and all other distances are in nautical miles. Time is in universal coordinated time (UTC).

Table 3-1 PIREP Format

PIREP Format	
UUA/UA	Type of report
OV	Location
TM	Time
FL	Altitude/Flight level
TP	Aircraft type
SK	Sky cover
WX	Flight visibility and weather
TA	Temperature
WV	Wind
TB	Turbulence
IC	Icing
RM	Remarks

Table 3-2 Encoding PIREPs

UUA/UA	Type of report:
	    URGENT (UUA) - Any PIREP that contains any of the following weather phenomena: tornadoes, funnel clouds, or waterspouts; severe or extreme turbulence, including clear air turbulence (CAT); severe icing; hail; volcanic ash: low-level wind shear (LLWS) (pilot reports air speed fluctuations of 10 knots or more within 2,000 feet of the surface); any other weather phenomena reported which are considered by the controller to be hazardous,
	    or potentially hazardous, to flight operations.
	    ROUTINE (UA) - Any PIREP that contains weather phenomena not listed above, including low-level wind shear reports with air speed fluctuations of less than 10 knots.

/OV	Location: Use VHF NAVAID(s) or an airport using the three- or four-letter location identifier. Position can be over a site, at some location relative to a site, or along a route. Ex: /OV ABC; /OV KFSM090025; /OV OKC045020-DFW; /OV KABR-KFSD
/TM	Time: Four digits in UTC. Ex: /TM 0915
/FL	Altitude/Flight level: Three digits for hundreds of feet with no space between FL and altitude. If not known, use UNKN. Ex: /FL095; /FL310; /FLUNKN
/TP	Aircraft type: Four digits maximum; if not known, use UNKN. Ex: /TP L329; /TP B737; /TP UNKN
/SK	Sky cover: Describes cloud amount, height of cloud bases, and height of cloud tops. If unknown, use UNKN. Ex: /SK SCT040-TOP080; /SK BKNUNKN-TOP075; /SK BKN-OVC050-TOPUNKN; /SK SCT030-TOP060/OVC120; /SK FEW030; /SK SKC
/WX	Flight visibility and weather: Flight visibility (FV) reported first in standard METAR weather symbols. Intensity (- for light, no qualifier for moderate, and + for heavy) shall be coded for all precipitation types except ice crystals and hail. Ex: /WX FV05SM -RA; /WX FV01SM SN BR; /WX RA
/TA	Temperature (Celsius): If below zero, prefix with an "M." Temperature shall also be reported if icing is reported. Ex: /TA 15; /TA M06
/WV	Wind: Direction from which the wind is blowing coded in tens of degrees using three digits. Directions of less than 100 degrees shall be preceded by a zero. The wind speed shall be entered as a two- or three-digit group immediately following the direction, coded in whole knots using the hundreds, tens, and units digits. Ex: /WV 27045KT; /WV 280110KT
/TB	Turbulence: Use standard contractions for intensity and type (CAT or CHOP when appropriate). Include altitude only if different from FL. (See Table 3-3.) Ex: /TB EXTRM; /TB OCNL LGT-MOD BLW 090; /TB MOD-SEV CHOP 080-110
/IC	Icing: Describe using standard intensity and type contractions. Include altitude only if different from FL. (See Table 3-4.) Ex: /IC LGT-MOD RIME; /IC SEV CLR 028-045
/RM	Remarks: Use free form to clarify the report putting hazardous elements first. Ex: /RM LLWS -15 KT SFC-030 DURC RWY22 JFK

Icing and turbulence reports state intensities using standard terminology when possible. To lessen the chance of misinterpretation, report icing and turbulence in standard terminology. If a PIREP stated,

"...PRETTY ROUGH AT 6,500, SMOOTH AT 8,500 PA24...," there could be many interpretations of the strength of the turbulence at 6,500 feet. A report of "light," "moderate," or "severe" turbulence at 6,500 feet would have been more concise and understandable. If a pilot's description of an icing or turbulence encounter cannot readily be translated into standard terminology, the pilot's description should be transmitted verbatim.

TURBULENCE

The following table classifies each turbulence intensity according to its effect on aircraft control, structural integrity, and articles and occupants within the aircraft.

Pilots should report location(s), time (UTC), altitude, aircraft type, whether in or near clouds, intensity, and when applicable, type (CHOP/clear air turbulence [CAT]), and duration of turbulence. Duration may be based on the time the pilot is flying between two locations or over a single location.

High-level turbulence (normally above 15,000 feet AGL) that is not associated with cumuliform clouds (including thunderstorms) shall be reported as CAT.

Table 3-3 Turbulence Reporting Criteria

Intensity	Aircraft Reaction	Reaction Inside Aircraft
Light	Turbulence that momentarily causes slight, erratic changes in altitude and/or attitude (pitch, roll, yaw). Report as light turbulence or light CAT. or Turbulence that causes slight, rapid and somewhat rhythmic bumpiness without appreciable changes in altitude or attitude. Report as light CHOP.	Occupants may feel a slight strain against belts or shoulder straps. Unsecured objects may be displaced slightly. Food service may be conducted and little or no difficulty is encountered in walking.
Moderate	Turbulence that causes changes in altitude and/or attitude occurs but the aircraft remains in positive control at all times. It usually causes variations in indicated airspeed. Report as moderate turbulence or moderate CAT. or Turbulence that is similar to light CHOP but of greater intensity. It causes rapid bumps or jolts without appreciable changes in aircraft or attitude. Report as moderate CHOP.	Occupants feel definite strains against seat belts or shoulder straps. Unsecured objects are dislodged. Food service and walking are difficult.
Severe	Turbulence that causes large, abrupt changes in altitude and/or attitude. It usually causes large variations in indicated airspeed. Aircraft may be momentarily out of control. Report as severe turbulence or severe CAT.	Occupants are forced violently against seat belts or shoulder straps. Unsecured objects are tossed about. Food service and walking are impossible.
Extreme	Turbulence in which the aircraft is violently tossed about and is practically impossible to control. It may cause structural damage. Report as extreme turbulence or extreme CAT.	

ICING

The following table classifies each icing intensity according to its operational effects on aircraft.

Pilots should report location(s), time (UTC), altitude, aircraft type, temperature, and icing intensity and type (rime, clear, or mixed). Rime ice is rough, milky, opaque ice formed by the instantaneous freezing of small supercooled water droplets. Clear ice is a glossy, clear, or translucent ice formed by the relatively slow freezing of large supercooled water droplets. Mixed ice is a combination of rime and clear ice.

Table 3-4 Icing Intensities, Airframe Ice Accumulation, and Pilot Report

Intensity	Airframe Ice Accumulation	Pilot Report
Trace	Ice becomes perceptible. Rate of accumulation slightly greater than rate of sublimation. It is not hazardous even though deicing/anti-icing equipment is not used unless encountered for an extended period of time (over 1 hour).	Location, time, altitude/FL, aircraft type, temperature, and icing intensity and type
Light	The rate of accumulation may create a problem if flight is prolonged in this environment (over 1 hour). Occasional use of deicing/anti-icing equipment removes/prevents accumulation. It does not present a problem if the deicing/anti-icing equipment is used.	Location, time, altitude/FL, aircraft type, temperature, and icing intensity and type
Moderate	The rate of accumulation is such that even short encounters become potentially hazardous and use of deicing/anti-icing equipment or diversion is necessary.	Location, time, altitude/FL, aircraft type, temperature, and icing intensity and type
Severe	The rate of accumulation is such that deicing/anti-icing equipment fails to reduce or control the hazard. Immediate diversion is necessary.	Location, time, altitude/FL, aircraft type, temperature, and icing intensity and type

EXAMPLES AND EXPLANATIONS (REFER TO TABLE 3-2):

UUA /OV ORD/TM 1235/FLUNKN/TP B727/TB MOD/RM LLWS +/- 20KT BLW 003 DURD RWY27L

Urgent UA over Chicago O'Hare Airport, Chicago, IL, at 1235Z. Flight level is unknown but the information is from a Boeing 727. Turbulence was moderate and on descent to runway 27 left, low-level wind shear was detected below 300 feet. Airspeed fluctuations were plus and minus 20 knots.

UUA /OV ABQ090045/TM 1430/FL130/TP BE30/TB SEV/RM BROKE ALL THE BOTTLES IN THE BAR

An urgent UA 45 miles east of Albuquerque, NM, a pilot of a Beech King Air 300 reported severe turbulence at 13,000 feet. The pilot remarked the turbulence was so severe it broke all the bottles in the passenger cabin bar.

UA /OV KMRB-KPIT/TM 1600/FL100/TP BE55/SK BKN024-TOP032/BKN-OVC043-TOPUNKN /TA M12/IC LGT-MOD RIME 055-080

This PIREP is decoded as follows: UA, Martinsburg to Pittsburgh, Pennsylvania (PA), at 1600 UTC at 10,000 feet MSL. Type of aircraft is a Beechcraft Baron. First cloud layer is broken with a base at 2,400 feet MSL broken and tops at 3,200 feet MSL. The second cloud layer is broken to occasionally overcast with a base at 4,300 feet MSL, and tops unknown. Outside air temperature is -12 degrees Celsius. Light to moderate rime icing is reported between 5,500 and 8,000 feet MSL.

UA /OV KOKC090064/TM 1522/FL080/TP C172/SK SCT090-TOPUNKN/WX FV05SM HZ/TA M04/WV 24540KT/TB LGT/RM IN CLR.

This PIREP is decoded as follows: UA, 64 nautical miles east of Oklahoma City VOR at 1522 UTC, flight level 8,000 feet MSL. Type of aircraft is a Cessna 172. There is a scattered cloud layer with bases at 9,000 feet MSL and unknown tops. Flight visibility is restricted to 5 statute miles due to haze. Outside air temperature is -4 degrees Celsius, wind is 245 degrees at 40 knots, light turbulence, and the aircraft is in clear skies.

UA /OV KLIT-KFSM/TM 0310/FL100/TP BE36/SK SCT070-TOP110/TA M03/WV 25015KT.

This PIREP is decoded as follows: UA between Little Rock and Fort Smith, Arkansas (AR), at 0310 UTC. A Beech 36 is at 10,000 feet MSL. There is a scattered cloud layer with bases at 7,000 feet MSL, and tops at 11,000 feet MSL. The outside air temperature is -3 degrees Celsius. Winds are from 250 degrees at 15 knots.

UA /OV KABQ/TM 1845/RM TIJERAS PASS CLSD DUE TO FG AND LOW CLDS UNA VFR RTN ABQ.

The PIREP is over Albuquerque at 1845 UTC. The remark section indicates the Tijeras pass is closed due to fog and low clouds. The pilot also mentions that she/he could not continue VFR and returned to Albuquerque.

UA /OV KTOL/TM 2200/FL310/TP B737/TB MOD CAT 350-390.

This PIREP is decoded as follows: UA over Toledo, Ohio, at 2200 UTC and flight level 310, a Boeing 737 reported moderate clear air turbulence between 35,000 and 39,000 feet MSL.

Nonmeteorological PIREPs sometimes help air traffic controllers. This "plain language" report stated:

… /RM 3N PNS LARGE FLOCK OF BIRDS HDG GEN N MAY BE SEAGULLS FRMN …

This PIREP alerted pilots and controllers to a bird hazard.

RADAR WEATHER REPORT (SD)

General areas of precipitation, including rain, snow, and thunderstorms, can be observed by radar. The radar weather report (SD) includes the type, intensity, and location of the echo top of the precipitation. (The intensity trend of precipitation is no longer coded on the SD.) It is important to remember that all heights are reported above MSL. Table 3-5 explains symbols denoting intensity. Radar stations report each hour at H+35.

Example of an SD:

TLX 1935	LN	8	TRW++	86/40 164/60	20W	C2425	MTS 570 AT 159/65	AUTO
a.	b.	c.	d.	e.	f.	g.	h.	i.

^MO1 NO2 ON3 PM34 QM3 RL2 =
           j.

Above SD report decoded as follows:

a. Location identifier and time of radar observation (Oklahoma City SD at 1935 UTC).
b. Echo pattern (LN in this example). The echo pattern or configuration may be one of the following:
    1. Line (LN) is a line of convective echoes with precipitation intensities that are heavy or greater, at least 30 miles long, at least 4 times as long as it is wide, and at least 25% coverage within the line.
    2. Area (AREA) is a group of echoes of similar type and not classified as a line.
    3. Cell (CELL) is a single isolated convective echo such as a rain shower.
c. Coverage, in tenths, of precipitation in the defined area (8/10 in this example).
d. Type and intensity of weather (thunderstorm [T] with very heavy rainshowers [RW++]).

Table 3-5 Precipitation Intensity

Symbol	Intensity
-	Light
(none)	Moderate
+	Heavy
++	Very Heavy
X	Intense
XX	Extreme

Table 3-6 Symbols Used in SD

Symbol	Meaning
R	Rain
RW	Rain shower
S	Snow
SW	Snow shower
T	Thunderstorm

Example of an SD:

TLX 1935	LN	8	TRW++	86/40 164/60	20W	C2425	MTS 570 AT 159/65	AUTO
a.	b.	c.	d.	e.	f.	g.	h.	i.

^MO1 NO2 ON3 PM34 QM3 RL2 =
    j.

e. Azimuth, referenced to true north, and range, in nautical miles (NM) from the radar site, of points defining the echo pattern (86/40 164/60 in this echo). For lines and areas, there will be two azimuth and range sets that define the pattern. For cells, there will be only one azimuth and range set. (See the examples that follow for elaboration of echo patterns.)
f. Dimension of echo pattern (20 NM wide in this example). The dimension of an echo pattern is given when azimuth and range define only the center line of the pattern. (In this example, "20W" means the line has a total width of 20 NM, 10 miles either side of a center line drawn from the points given in item "e" above.)
g. Cell movement (cells within line moving from 240 degrees at 25 knots in this example). Movement is only coded for cells; it will not be coded for lines or areas.
h. Maximum top and location (57,000 feet MSL on radial 159 degrees at 65 NM in this example). Maximum tops may be coded with the symbols "MT" or "MTS." If it is coded with "MTS" it means that satellite data as well as radar information was used to measure the top of the precipitation.
i. The report is automated from WSR-88D weather radar data.
j. Digital section is used for preparing radar summary chart.

To aid in interpreting SDs, the five following examples are decoded into plain language.

GRB 1135 AREA 4TRW+ 9/100 130/75 50W C2425 MT 310 at 45/47 AUTO

Green Bay, WI, Automated SD at 1135 UTC. An area of echoes, 4/10 coverage, containing thunderstorms and heavy rain showers. Area is defined by points (referenced from GRB radar site) at 9 degrees, 100 NM and 130 degrees, 75 NM. These points, plotted on a map and connected with a straight line, define the center line of the echo pattern. The width of the area is 50 NM; i.e., 25 NM either side of the center line. The cells are moving from 240 degrees at 25 knots. Maximum top is 31,000 feet MSL located at 45 degrees and 47 NM from GRB.

ICT 1935 LN 9TRWX 275/80 210/90 20W C2430 MTS 440 AT 260/48 AUTO

Wichita, KS, Automated SD at 1935 UTC. A line of echoes, 9/10 coverage, containing thunderstorm with intense rain showers. The center of the line extends from 275 degrees, 80 NM to 210 degrees, 90 NM. The line is 20 NM wide. NOTE: To display graphically, plot the center points on a map and connect the points with a straight line; then plot the width. Since the thunderstorm line is 20 miles wide, it extends 10 miles either side of your plotted line. The thunderstorm cells are moving from 240 degrees at 30 knots. The maximum top is 44,000 feet MSL at 260 degrees, 48 NM from ICT.

GGW 1135 AREA 3S- 90/120 150/80 34W MT 100 at 130/49

Glasgow, MT, Automated SD at 1135 UTC. An area, 3/10 coverage, of light snow. The area's centerline extends from points at 90 degrees, 120 NM to 150 degrees, 80 NM. The area is 34 NM wide. No movement was reported. The maximum top is 10,000 feet MSL, at 130 degrees, 49 NM.

MAF 1135 AREA 2TRW++6R- 67/130 308/45 105W C2240 MT 380 AT 66/54

Midland/Odessa, TX, Automated SD at 1135 UTC. An area of echoes, total coverage 8/10, with 2/10 of thunderstorms with very heavy rainshowers and 6/10 coverage of light rain. (This suggests that the thunderstorms are embedded in an area of light rain.) The area lies 52½ miles either side of the line defined by the two points, 67 degrees, 130 NM and 308 degrees, 45 NM.

When an SD is transmitted but does not contain any encoded weather observation, a contraction is sent which indicates the operational status of the radar.

Example:
TLX 1135 PPINE AUTO

It is decoded as Oklahoma City, OK's, radar at 1135 UTC detects no echoes.

Table 3-7 Operational Status Contractions

Contraction	Operational Status
PPINE	Radar is operating normally but there are no echoes being detected.
PPINA	Radar observation is not available.
PPIOM	Radar is inoperative or out of service.
AUTO	Automated radar report from WSR-88D.

All SDs also contain groups of digits.

Example:
^MO1 NO1 ON3 PM34 QM3 RL2 SL1=

These groups of digits are the final entry on the SD. This digitized radar information is used primarily in preparing the radar summary chart. However, by using a proper grid overlay chart for the corresponding radar site, this code is also useful in determining more precisely where the precipitation is occurring within an area as well as the intensity of the precipitation. (See Figure 3-1 for an example of a digital code plotted from the Oklahoma City, OK, SD.)

The digit assigned to a box represents the intensity of precipitation as determined by the WSR-88D and is the <u>maximum</u> precipitation intensity found within the grid box. (See Table 7-2 for definitions of precipitation intensities associated with digits 1 through 6.) These digits were once commonly referred to as VIP levels because precipitation intensity, and therefore the digits, was derived using a video integrator processor (VIP). Since the WSR-88D and <u>not</u> the video integrator processor is now used to determine precipitation intensity, it is suggested that the term VIP should <u>no longer</u> be used when describing precipitation intensity. For example, if a specific grid has the number 2 associated with it, that grid would be described as having moderate precipitation, not VIP level 2 precipitation.

A box is identified by two letters, the first representing the row in which the box is found and the second letter representing the column. For example "MO1" identifies the box located in row M and column O as containing light precipitation. A code of "MO1234" indicates precipitation in four consecutive boxes in the same row. Working from left to right box MO = 1, box MP = 2, MQ = 3 and box MR = 4.

When using hourly SDs in preflight planning, note the location and coverage of echoes, the type of weather reported, the intensity, and especially the direction of movement.

It is important to remember that the SD contains information pertaining to the location of particles in the atmosphere that are of precipitation size or larger. It does <u>not</u> display locations of cloud-size particles, and, therefore, neither ceilings nor restrictions to visibility. An area may be blanketed with fog or low stratus, but the SD would not include information about it. Pilots should use SDs along with METARs, satellite photos, and forecasts when planning a flight.

The SDs help pilots plan ahead to avoid thunderstorm areas. Once airborne, however, pilots must depend on contact with Flight Watch, which has the capability to display current radar images, airborne radar, or visual sighting to evade individual storms.

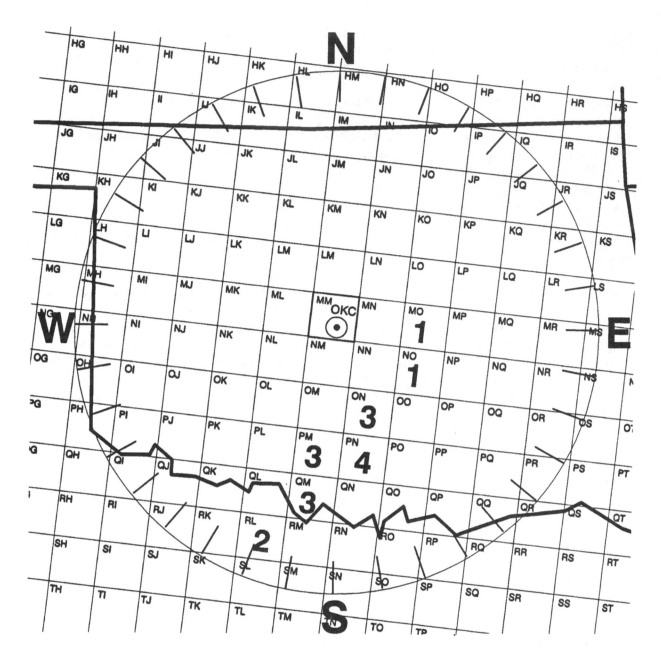

Figure 3-1. Digital Radar Report Plotted on a PPI Grid Overlay Chart.
(Note: See Table 7-2 for Intensity Level Codes 1 through 6.)

SATELLITE WEATHER PICTURES

Prior to weather satellites, weather observations were made only at distinct points within the atmosphere and supplemented by PIREPs. These PIREPs gave a "sense" of weather as viewed from above. However, with the advent of weather satellites, a whole new dimension to weather observing and reporting has emerged. There are two types of weather satellites in use by the U. S. today: Geostationary Operational Environmental Satellite (GOES), which is a geostationary satellite, and the Polar Orbiter Environmental Satellite (POES). Additional satellite imagery is available from the European Meteosat and the Japanese GMS geostationary satellites.

Two U.S. GOES satellites are used for imaging. One is stationed over the equator at 75 degrees west longitude and is referred to as GOES EAST since it covers the eastern U.S. The other is positioned at 135 degrees west longitude and is referred to as GOES WEST since it covers the western U.S. Together they cover North and South America and surrounding waters. They normally transmit an image of Earth, pole to pole, every 15 minutes. When disastrous weather threatens the U.S., the satellites can scan small areas rapidly so that a picture can be received as often as every 1 minute. Data from these rapid scans are used at NWS offices.

Since the GOES satellite is stationary over the equator, the images poleward of about 50 degrees latitude become greatly distorted. For images above 50 degrees latitude, polar orbiting satellites are employed. The NOAA satellite is a polar orbiter and orbits the earth on a track that nearly crosses the North and South poles. A high resolution picture is produced about 500 miles either side of its track on the journey from pole to pole. The NOAA pictures are essential to weather personnel in Alaska and Canada.

Two types of imagery are available from satellites, and, when combined, give a great deal of information about clouds. Through interpretation, the analyst can determine the type of cloud, the temperature of cloud tops (from this, the approximate height of the cloud can be determined), and the thickness of cloud layers. From this information, the analyst gets a good idea of the associated weather.

One type of imagery is visible (Figure 3-2). A visible image shows clouds and Earth reflecting sunlight to the satellite sensor. The greater the reflected sunlight reaching the sensor, the whiter the object is on the picture. The amount of reflectivity reaching the sensor depends upon the height, thickness, and ability of the object to reflect sunlight. Since clouds are much more reflective than most of the Earth, clouds will usually show up white on the picture, especially thick clouds. Thus, the visible picture is primarily used to determine the presence of clouds and the type of cloud from shape and texture. Due to the obvious lack of sunlight, there are no visible images available at night.

The second type of imagery is infrared (IR) (Figure 3-3). An IR picture shows heat radiation being emitted by clouds and Earth. The images show temperature differences between cloud tops and the ground, as well as temperature gradations of cloud tops and along the Earth's surface. Ordinarily, cold temperatures are displayed as light gray or white. High clouds appear the whitest. However, various computer-generated enhancements are sometimes used to sharply illustrate important temperature contrasts. IR images measure cloud top temperatures and are used to approximate the height of clouds. From this, one can see the importance of using visible and IR imagery together when interpreting clouds. IR images are available both day and night.

Satellite images are processed by the NWS as well as by many private companies. Therefore, they can be received from many different sources. Depending upon the source, satellite images may be updated anywhere from every 15 minutes to every hour; therefore, it is important to note the time on the images when interpreting them. By viewing satellite images, the development and dissipation of weather can be seen and followed over the entire country and coastal regions.

NESDIS is developing the capability to provide derived products useful to aviation from satellite data. These experimental products are available via the Internet and include:

1. Fog and low cloud coverage and depth.
2. Volcanic ash detection.
3. Microburst products.
4. Soundings.
5. Clear air turbulence.
6. Aircraft icing potential.

Figure 3-2. Visible Satellite Imagery.

Figure 3-3. Infrared Satellite Imagery.

RADIOSONDE ADDITIONAL DATA (RADATs)

Radiosonde Additional Data (RADATs) information is obtained from the radiosonde observations that are conducted twice a day at 00 and 12Z. The information contained in a RADAT is the observed freezing level and the relative humidity associated with the freezing level. The freezing level is the height above MSL at which the temperature is zero degrees Celsius.

The format associated with the RADAT is as follows:
Stn ID Time RADAT UU (D) (hhh)(hhh)(hhh)(/n)

Explanation:
Stn ID and Time - standard three-letter identifier and observation time in UTC.

RADAT - a contraction identifying the data as "freezing-level data."

UU - relative humidity at the freezing level in percent. When more than one level is identified, "UU" is the highest relative humidity observed at any of the levels transmitted.

(D) - a coded letter "L," "M," or "H." used in the event of multiple freezing levels to identify which level has the highest relative humidity, "L – lowest," "M – middle," "H – highest." This letter is omitted when only one level is coded.

(hhh) – height of the freezing level in hundreds of feet. Up to three freezing levels can be specified in the event of multiple freezing levels. If there are more than three freezing levels, the levels coded are the lowest, highest, and the intermediate level with the highest relative humidity.

(/n) – an indicator to show the number of freezing levels in addition to the three which are coded. The number is omitted when all observed freezing levels are coded (three or less.)

Examples:
SJU 1200 RADAT 87160
The San Juan, Puerto Rico, RADAT indicates that the freezing level was 16,000 feet MSL and the relative humidity was 87% at the freezing level.

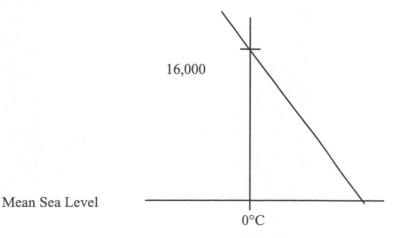

16,000

Mean Sea Level

0°C

Figure 3-4. SJU RADAT.

OUN 0000 RADAT 87L024105
The Norman, Oklahoma, RADAT indicates that the freezing level was crossed twice. The two crossings occurred at 2,400 feet MSL and at 10,500 feet MSL. The 87L indicates that the relative humidity was 87% at the lowest crossing (indicated by the L).

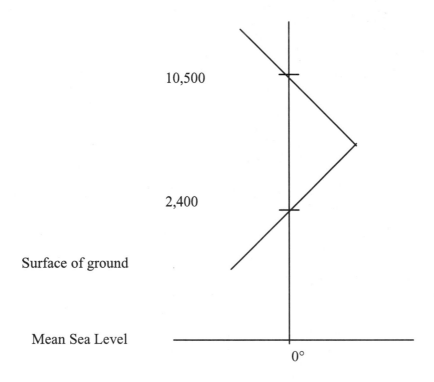

Figure 3-5. OUN RADAT.

ALB 1200 RADAT 84M019045051

The Albany, New York, RADAT indicates three crossings of the freezing level. The three crossings of the zero-degree Celsius isotherm occurred at 1,900 feet MSL, 4,500 feet MSL, and at 5,100 feet MSL. The relative humidity was 84% at the middle crossing which was 4,500 feet MSL.

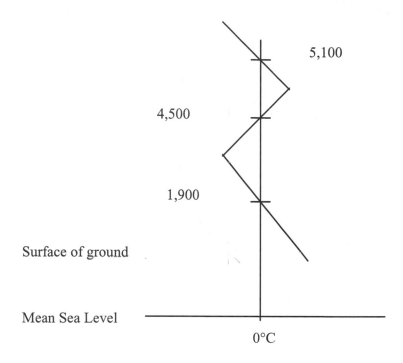

Figure 3-6. ALB RADAT.

DNR 1200 RADAT ZERO

The Denver, Colorado, RADAT indicates that the entire RADAT information was below zero degrees Celsius.

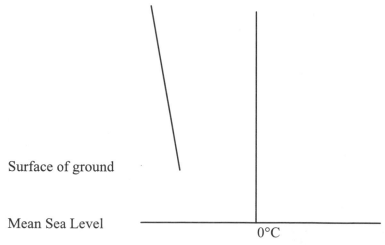

Figure 3-7. DEN RADAT.

ABR 0000 RADAT MISG
The Aberdeen, South Dakota, RADAT was terminated before the first crossing of the zero-degree Celsius isotherm. All temperatures were above freezing.

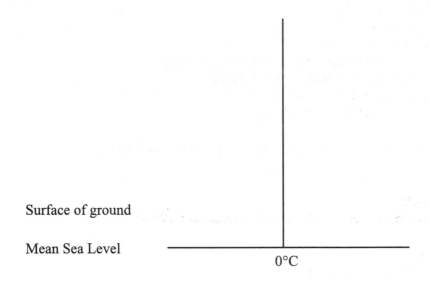

Figure 3-8. ABR RADAT.

Section 4
AVIATION WEATHER FORECASTS

Good flight planning involves considering all available weather information, including weather forecasts. This section explains the following aviation forecasts:

1. Aviation Terminal Forecast (TAF)
2. Aviation Area Forecast (FA)
3. Inflight Aviation Weather Advisories
4. Alaska, Gulf of Mexico, and International Area Forecasts (FAs)
5. Transcribed Weather Broadcasts (TWEB) Text Products
6. Winds and Temperatures Aloft Forecast (FD)
7. Center Weather Service Unit (CWSU) Products

Also discussed are the following general forecasts that may aid in flight planning:

1. Hurricane Advisory (WH)
2. Convective Outlook (AC)
3. Severe Weather Watch Bulletins (WW) and Alert Messages (AWW)

AVIATION TERMINAL FORECAST (TAF)

An Aviation Terminal Forecast (TAF) is a concise statement of the expected meteorological conditions within a 5-statute-mile radius from the center of an airport's runway complex during a 24-hour time period.

The TAFs use the same weather code found in METAR weather reports. Detailed explanations of the code are found only in Section 2.

The National Weather Service (NWS) requires an airport to have two consecutive METAR observations, not less than 30 minutes apart nor more than 1 hour apart, before a TAF will be issued. After the TAF has been issued, the forecaster will use all available weather data sources to maintain the TAF. If during this time a METAR is missing or part of the METAR is missing, the forecaster can use other weather sources to obtain the necessary data to maintain the TAF. However, if the forecaster feels that the other weather sources cannot provide the necessary information, the forecaster will discontinue the TAF.

A TAF contains the following elements in the order listed:

1. Type of report
2. ICAO station identifier
3. Date and time of origin
4. Valid period date and time
5. Wind forecast
6. Visibility forecast
7. Significant weather forecast
8. Sky condition forecast
9. Nonconvective low-level wind shear forecast (optional data)
10. Forecast change indicators
11. Probability forecast

International and U.S. military TAFs also contain forecasts of maximum and minimum temperature, icing, and turbulence. These three elements are <u>not</u> included in NWS-prepared TAFs. For forecast icing and turbulence, see page 4-23, Inflight Aviation Weather Advisories.

The following paragraphs describe the elements in a TAF report. A sample report will accompany each element with the subject element in bold letters.

TYPE OF REPORT

TAF
KPIR 111140Z 111212 13012KT P6SM BKN100 WS020/35035KT TEMPO 1214 5SM BR
  FM1500 16015G25KT P6SM SCT040 BKN250
  FM0000 14012KT P6SM BKN080 OVC150 PROB40 0004 3SM TSRA BKN030CB
  FM0400 14008KT P6SM SCT040 OVC080 TEMPO 0408 3SM TSRA OVC030CB
   BECMG 0810 32007KT=

The report type header will always appear as the first element in the TAF. There are two types of TAF reports: a routine forecast, **TAF**; and an amended forecast, **TAF AMD**. An amended TAF is issued when the forecaster feels the TAF is not representative of the current or expected weather conditions. An equal sign at the end of the TAF signifies the end of the report.

ICAO STATION IDENTIFIER

TAF
KPIR 111140Z 111212 13012KT P6SM BKN100 WS020/35035KT TEMPO 1214 5SM BR
  FM1500 16015G25KT P6SM SCT040 BKN250
  FM0000 14012KT P6SM BKN080 OVC150 PROB40 0004 3SM TSRA BKN030CB
  FM0400 14008KT P6SM SCT040 OVC080 TEMPO 0408 3SM TSRA OVC030CB
   BECMG 0810 32007KT=

The TAF code uses ICAO four-letter location identifiers as described in Section 2. TAF locations are in Figures 4-1, 4-2, 4-3, and 4-4 located on pages 4-13 through 4-16.

DATE AND TIME OF ORIGIN

TAF
KPIR **111140Z** 111212 13012KT P6SM BKN100 WS020/35035KT TEMPO 1214 5SM BR
  FM1500 16015G25KT P6SM SCT040 BKN250
  FM0000 14012KT P6SM BKN080 OVC150 PROB40 0004 3SM TSRA BKN030CB
  FM0400 14008KT P6SM SCT040 OVC080 TEMPO 0408 3SM TSRA OVC030CB
   BECMG 0810 32007KT=

This element is the date and universal coordinated time (UTC) the forecast is actually prepared. The format is a two-digit date and four-digit time followed without a space by the letter **Z**. Routine TAFs are prepared and filed approximately one-half hour prior to scheduled issuance times.

Examples:
111140Z Forecast prepared on the eleventh day of the month at 1140Z.
050530Z Forecast prepared on the fifth day of the month at 0530Z.

VALID PERIOD DATE AND TIME

TAF

KPIR 111140Z **111212** 13012KT P6SM BKN100 WS020/35035KT TEMPO 1214 5SM BR
   FM1500 16015G25KT P6SM SCT040 BKN250
   FM0000 14012KT P6SM BKN080 OVC150 PROB40 0004 3SM TSRA BKN030CB
   FM0400 14008KT P6SM SCT040 OVC080 TEMPO 0408 3SM TSRA OVC030CB
   BECMG 0810 32007KT=

The valid period of the forecast is a two-digit date followed by the two-digit beginning and two-digit ending hours in UTC. Routine TAFs are valid for 24 hours and are issued four times daily at 0000Z, 0600Z, 1200Z, and 1800Z. All ending times throughout the TAF of 00Z are indicated by the number 24.

Examples:
111212 Forecast valid from the eleventh at 12Z to the twelfth at 12Z.
300024 Forecast valid from the thirtieth at 00Z to the first at 00Z.

Amended, canceled, or delayed forecasts may have valid periods less than 24 hours.

Examples:
231512 Forecast valid from the twenty-third at 15Z to the twenty-fourth at 12Z.
091006 Forecast valid from the ninth at 10Z to the tenth at 06Z.

For airports with less than 24-hour observational coverage for which part-time terminal forecasts are provided, the TAF will be valid until the end of the scheduled forecast even if the observations have ceased before that time. **AMD NOT SKED** (amendment not scheduled) or **NIL AMD** (no amendment) will be issued after the forecast information. **AMD NOT SKED AFT (closing time)Z** (amendment not scheduled after [closing time]Z) will be used if the times of the observations are known and judged reliable. During the time the station is closed and a TAF is issued, there will be no forecast as indicated by **NIL** (no TAF) after the valid date and time group. Only after two METARs observations have been disseminated will a TAF be issued. **AMD LTD TO CLD VIS AND WIND** (amendment limited to clouds, visibility, and wind) is used at observation sites that have part-time manual augmentation. This remark means that there will be amendments only for clouds, visibility, and wind. There will be <u>no</u> amendments for thunderstorms or freezing/frozen precipitation.

WIND FORECAST

TAF

KPIR 111140Z 111212 **13012KT** P6SM BKN100 WS020/35035KT TEMPO 1214 5SM BR
   FM1500 16015G25KT P6SM SCT040 BKN250
   FM0000 14012KT P6SM BKN080 OVC150 PROB40 0004 3SM TSRA BKN030CB
   FM0400 14008KT P6SM SCT040 OVC080 TEMPO 0408 3SM TSRA OVC030CB
   BECMG 0810 32007KT=

The surface wind forecast is the wind direction in degrees from true north (first three digits) and mean speed in knots (last two or three digits if 100 knots or greater). The contraction, **KT**, denotes the units of wind speed in knots. Wind gusts are noted by the letter **G** appended to the mean wind speed followed by the highest expected gust (two or three digits if 100 knots or greater). Calm winds are encoded as **00000KT**. A variable wind is encoded as **VRB** when wind direction fluctuates due to convective activity or low wind speeds (3 knots or less).

Examples:
13012KT, 18010KT, 35012G26KT, or VRB16G28KT

VISIBILITY FORECAST

TAF
KPIR 111140Z 111212 13012KT **P6SM** BKN100 WS020/35035KT TEMPO 1214 5SM BR
   FM1500 16015G25KT P6SM SCT040 BKN250
   FM0000 14012KT P6SM BKN080 OVC150 PROB40 0004 3SM TSRA BKN030CB
   FM0400 14008KT P6SM SCT040 OVC080 TEMPO 0408 3SM TSRA OVC030CB
   BECMG 0810 32007KT=

The prevailing visibility is forecasted in whole and fractions of statute miles followed by **SM** to note the units of measurement. Statute miles followed by fractions of statute miles are separated with a space; for example, 1 1/2SM. Forecasted visibility greater than 6 statute miles is indicated by coding **P6SM**. If prevailing visibility is 6 statute miles or less, one or more weather phenomena must be included in the significant weather forecast. If volcanic ash is forecasted, the visibility must also be forecasted even if the visibility is greater than 6 statute miles. Sector or variable visibility is <u>not</u> forecasted.

Examples:
1/2SM, 2 1/4SM, 5SM, or P6SM

SIGNIFICANT WEATHER FORECAST

TAF
KPIR 111140Z 111212 13012KT P6SM BKN100 WS020/35035KT TEMPO 1214 5SM BR
   FM1500 16015G25KT P6SM SCT040 BKN250
   FM0000 14012KT P6SM BKN080 OVC150 PROB40 0004 3SM **TSRA** BKN030CB
   FM0400 14008KT P6SM SCT040 OVC080 TEMPO 0408 3SM TSRA OVC030CB
   BECMG 0810 32007KT=

The expected weather phenomenon or phenomena are coded in TAF reports using the same format, qualifiers, and phenomena contractions as METAR reports (except UP). (See Section 2.)

Obscurations to vision will be forecasted whenever the prevailing visibility is forecasted to be 6 statute miles or less. Precipitation and volcanic ash will always be included in the TAF regardless of the visibility forecasted.

Examples:
FM2200 18005KT 1SM BR SKC
FM0100 12010KT P6SM -RA BKN020
FM1500 22015KT P6SM VA SCT100

If no significant weather is expected to occur during a specific time period in the forecast, the weather group is omitted for that time period. However, if after a time period in which significant weather has been forecasted, a change to a forecast of "no significant weather" occurs, the contraction **NSW** (<u>n</u>o <u>s</u>ignificant <u>w</u>eather) will appear as the weather included in BECMG or TEMPO groups. NSW will not be used in the initial time period of a TAF or in FM groups.

Example:
FM0600 16010KT 3SM RA BKN030 BECMG 0810 P6SM NSW

If the forecaster determines that in the vicinity of the airport there could be weather that impacts aviation, the forecaster will include those conditions after the weather group. The letters **VC** describe conditions that will occur within the vicinity of an airport (5-10 SM) and will be used only with fog, showers, or thunderstorms (FG, SH, or TS).

Examples:
P6SM VCFG - fog in the vicinity.
5SM BR VCSH - showers in the vicinity .
P6SM VCTS - thunderstorms in the vicinity.

SKY CONDITION FORECAST

TAF
KPIR 111140Z 111212 13012KT P6SM **BKN100** WS020/35035KT TEMPO 1214 5SM BR
    FM1500 16015G25KT P6SM SCT040 BKN250
    FM0000 14012KT P6SM BKN080 OVC150 PROB40 0004 3SM TSRA BKN030CB
    FM0400 14008KT P6SM SCT040 OVC080 TEMPO 0408 3SM TSRA OVC030CB
    BECMG 0810 32007KT=

TAF sky condition forecasts use the METAR format described in Section 2. Cumulonimbus clouds (**CB**) are the only cloud type forecasted in TAFs.

Examples:
BKN100, SCT040 BKN030CB, or FEW008 BKN015

When the sky is obscured due to a surface-based phenomenon, vertical visibility (**VV**) into the obscuration is forecasted. The format for vertical visibility is **VV** followed by a three-digit height in hundreds of feet. Partial obscurations are not forecasted. Remember a ceiling is the lowest broken or overcast layer or vertical visibility.

Example:
VV008

NONCONVECTIVE LOW-LEVEL WIND SHEAR FORECAST (OPTIONAL DATA)

TAF
KPIR 111140Z 111212 13012KT P6SM BKN100 **WS020/35035KT** TEMPO 1214 5SM BR
    FM1500 16015G25KT P6SM SCT040 BKN250
    FM0000 14012KT P6SM BKN080 OVC150 PROB40 0004 3SM TSRA BKN030CB
    FM0400 14008KT P6SM SCT040 OVC080 TEMPO 0408 3SM TSRA OVC030CB
    BECMG 0810 32007KT=

A forecast of nonconvective low-level wind shear is included immediately after the cloud and obscuration group when wind shear criteria have been or will be met. The forecast includes the height of the wind shear followed by the wind direction and wind speed at the indicated height. Height is given in hundreds of feet above ground level (AGL) up to and including 2,000 feet. Wind shear is encoded with the contraction **WS**, followed by a three-digit height, solidus (/), and winds at the height indicated in the same format as surface winds. The wind shear element is omitted if not expected to occur.

Example:
WS020/36035KT

FORECAST CHANGE INDICATORS

TAF
KPIR 111140Z 111212 13012KT P6SM BKN100 WS020/35035KT **TEMPO 1214** 5SM BR
  FM1500 16015G25KT P6SM SCT040 BKN250
  FM0000 14012KT P6SM BKN080 OVC150 PROB40 0004 3SM TSRA BKN030CB
  FM0400 14008KT P6SM SCT040 OVC080 TEMPO 0408 3SM TSRA OVC030CB
  BECMG 0810 32007KT=

If a significant change in any of the elements is expected during the valid period, a new time period with the changes is included. The following change indicators are used when either a rapid, gradual, or temporary change is expected in some or all of the forecasted meteorological conditions.

From (FM) Group

The **FM** group is used when a <u>rapid</u> and significant change, usually occurring in less than 1 hour, in prevailing conditions is expected. Appended to the FM indicator is the <u>four-digit hour and minute</u> the change is expected to begin. The forecast is valid until the next change group or until the end of the current forecast.

The FM group will mark the beginning of a new line in a TAF report. Each FM group shall contain a forecast of wind, visibility, weather (if significant), sky condition, and wind shear (if warranted). FM groups will not include the contraction NSW.

Examples:
FM1500 16015G25KT P6SM SCT040 BKN250
FM0200 32010KT 3SM TSRA FEW010 BKN030CB

Becoming (BECMG) Group

The **BECMG** group is used when a <u>gradual</u> change in conditions is expected over a period not to exceed 2 hours. The time period when the change is expected to occur is a four-digit group containing the beginning and ending hours of the change that follows the BECMG indicator. The gradual change will occur at an unspecified time within the time period. Only the changing forecasted meteorological conditions are included in **BECMG** groups. Omitted conditions are carried over from the previous time group.

Example:
FM2000 18020KT P6SM BKN030 BECMG 0103 OVC015

This BECMG group describes a gradual change in sky condition from BKN030 to OVC015. The change in sky conditions occurs between 01Z and 03Z. Refer back to the FM2000 group for the wind and visibility conditions. The forecast after 03Z will be: 18020KT P6SM OVC015.

Example:
FM0400 14008KT P6SM SCT040 OVC080 TEMPO 0408 3SM TSRA OVC030CB
 BECMG 0810 32007KT=

This BECMG group describes a gradual change in wind direction only beginning between 08Z and 10Z. Refer back to the previous forecast group, in this case the FM0400 group, for the prevailing visibility, weather, and sky conditions. The forecast after 10Z will be: 32007KT P6SM SCT040 OVC080.

Temporary (TEMPO) Group

The **TEMPO** group is used for temporary fluctuations of wind, visibility, weather, or sky condition that are expected to last for generally less than an hour at a time (occasional), and expected to occur during less than half the time period. The **TEMPO** indicator is followed by a four-digit group giving the beginning and ending hours of the time period during which the temporary conditions are expected. Only the changing forecasted meteorological conditions are included in **TEMPO** groups. The omitted conditions are carried over from the previous time group.

Example:
FM1000 27005KT P6SM SKC TEMPO 1216 3SM BR

This temporary group describes visibility and weather between 12Z and 16Z. The winds and sky condition have been omitted. Go back to the previous forecast group, FM1000, to obtain the wind and sky condition forecast. The forecast between 12Z and 16Z is: 27005KT 3SM BR SKC.

Example:
FM0400 14008KT P6SM SCT040 OVC080 TEMPO 0408 3SM TSRA OVC030CB
 BECMG 0810 32007KT=

This temporary group describes visibility, weather, and sky condition between 04Z and 08Z. The winds have been omitted. Go back to the previous forecast group, FM0400, to obtain the wind forecast. The forecast between 04Z and 08Z is: 14008KT 3SM TSRA OVC030CB.

PROBABILITY (PROB30 or PROB40) FORECAST

TAF
KPIR 111140Z 111212 13012KT P6SM BKN100 WS020/35035KT TEMPO 1214 5SM BR
   FM1500 16015G25KT P6SM SCT040 BKN250
   FM0000 14012KT P6SM BKN080 OVC150 **PROB40 0004** 3SM TSRA BKN030CB
   FM0400 14008KT P6SM SCT040 OVC080 TEMPO 0408 3SM TSRA OVC030CB
    BECMG 0810 32007KT=

The probability forecast describes the probability or chance of thunderstorms or other precipitation events occurring, along with associated weather conditions (wind, visibility, and sky conditions). The probability forecast will not be used in the first 6 hours of the TAF.

The **PROB30** or **PROB40** group is used when the occurrence of thunderstorms or precipitation is in the 30% to less than 40% or 40% to less than 50% range, respectively. If the thunderstorms or precipitation chance is greater than 50%, it is considered a prevailing weather condition and is included in the significant weather section or the TEMPO change indicator group. PROB30 or PROB40 is followed by a four-digit time group giving the beginning and ending hours of the time period during which the thunderstorms or precipitation is expected.

Example:
FM0600 0915KT P6SM BKN020 PROB30 1014 1SM RA BKN015

This example depicts a 30% to less than 40% chance of 1statute mile, moderate rain, and a broken cloud layer (ceiling) at 1,500 feet between the hours of 10-14Z.

Example:
FM0000 14012KT P6SM BKN080 OVC150 PROB40 0004 3SM TSRA BKN030CB

In this example, there is a 40% to <50% chance of visibility 3 statute miles, thunderstorms with moderate rain showers, and a broken cloud layer (ceiling) at 3,000 feet with cumulonimbus between the hours of 00-04Z.

EXAMPLES OF TAF REPORTS

TAF
KPIR 111140Z 111212 13012KT P6SM BKN100 WS020/35035KT TEMPO 1214 5SM BR
   FM1500 16015G25KT P6SM SCT040 BKN250
      FM0000 14012KT P6SM BKN080 OVC150 PROB40 0004 3SM TSRA BKN030CB
      FM0400 14008KT P6SM SCT040 OVC080 TEMPO 0408 3SM TSRA OVC030CB
      BECMG 0810 32007KT=

TAF	Aviation terminal forecast
KPIR	Pierre, South Dakota
111140Z	prepared on the 11th at 1140Z
111212	valid period from the 11th at 1200Z until the 12th at 1200Z
13012KT	wind 130 at 12 knots
P6SM	visibility greater than 6 statute miles
BKN100	ceiling 10,000 broken
WS020/35035KT	wind shear at 2,000 feet, wind (at 2,000 feet) from 350 at 35 knots
TEMPO 1214	temporary conditions between 1200Z and 1400Z
5SM	visibility 5 statute miles
BR	mist
FM1500	from 1500Z
16015G25KT	wind 160 at 15 knots gusting to 25 knots
P6SM	visibility greater than 6 statute miles
SCT040 BKN250	4,000 scattered, ceiling 25,000 broken
FM0000	from 0000Z
14012KT	wind 140 at 12 knots
P6SM	visibility greater than 6 statute miles
BKN080 OVC150	ceiling 8,000 broken, 15,000 overcast
PROB40 0004	40% probability between 0000Z and 0400Z
3SM	visibility 3 statute miles
TSRA	thunderstorm with moderate rain showers
BKN030CB	ceiling 3,000 broken with cumulonimbus
FM0400	from 0400Z
14008KT	wind 140 at 8 knots
P6SM	visibility greater than 6 statute miles
SCT040 OVC080	4,000 scattered, ceiling 8,000 overcast
TEMPO 0408	temporary conditions between 0400Z and 0800Z
3SM	visibility 3 statute miles
TSRA	thunderstorms with moderate rain showers
OVC030CB	ceiling 3,000 overcast with cumulonimbus
BECMG 0810	becoming between 0800Z and 1000Z
32007KT=	wind 320 at 7 knots; the equal sign signifies the end of the TAF

TAF AMD
KEYW 131555Z 131612 VRB03KT P6SM VCTS SCT025CB BKN250 TEMPO 1618 2SM TSRA
   BKN020CB
    FM1800 VRB03KT P6SM SCT025 BKN250 TEMPO 2024 1SM TSRA OVC010CB
    FM0000 VRB03KT P6SM VCTS SCT020CB BKN120 TEMPO 0812 BKN020CB=

TAF AMD	Amended aviation terminal forecast
KEYW	Key West, Florida
131555Z	prepared on the 13th at 1555Z
131612	valid period from the 13th at 1600Z until the 14th at 1200Z
VRB03KT	wind variable at 3 knots
P6SM	visibility greater than 6 statute miles
VCTS	thunderstorms in the vicinity
SCT025CB BKN250	2,500 scattered with cumulonimbus, ceiling 25,000 broken
TEMPO 1618	temporary conditions between 1600Z and 1800Z
2SM	visibility 2 statute miles
TSRA	thunderstorms with moderate rain showers
BKN020CB	ceiling 2,000 broken with cumulonimbus
FM1800	from 1800Z
VRB03KT	wind variable at 3 knots
P6SM	visibility greater than 6 statute miles
SCT025 BKN250	2,500 scattered, ceiling 25,000 broken
TEMPO 2024	temporary conditions between 2000Z and 0000Z
1SM	visibility 1 statute mile
TSRA	thunderstorms with moderate rain showers
OVC010CB	ceiling 1,000 overcast with cumulonimbus
FM0000	from 0000Z
VRB03KT	variable wind at 3 knots
P6SM	visibility greater than 6 statute miles
VCTS	thunderstorms in the vicinity
SCT020CB BKN120	2,000 scattered with cumulonimbus, ceiling 12,000 broken
TEMPO 0812	temporary conditions between 0800Z and 1200Z
BKN020CB= the	ceiling 2,000 broken with cumulonimbus; the equal sign signifies the end of TAF

TAF
KCRP 111730Z 111818 19007KT P6SM SCT030 TEMPO 1820 BKN040
    FM2000 16011KT P6SM VCTS FEW030CB SCT250
    FM0200 14006KT P6SM FEW025 SCT250
    FM0800 VRB03KT 5SM BR SCT012 TEMPO 1012 1/2SM FG BKN001
    FM1500 17007KT P6SM SCT025=

TAF	Aviation terminal forecast
KCRP	Corpus Christi, Texas
111730Z	prepared on the 11th at 1730Z
111818	valid period from the 11th at 1800Z until the 12th at 1800Z
19007KT	wind 190 at 7 knots
P6SM	visibility greater than 6 statute miles
SCT030	3,000 scattered
TEMPO 1820	temporary conditions between 1800Z and 2000Z
BKN040	ceiling 4,000 broken
FM2000	from 2000Z
16011KT	wind 160 at 11 knots
P6SM	visibility greater than 6 statute miles
VCTS	thunderstorms in the vicinity
FEW030CB SCT250	3,000 few with cumulonimbus, 25,000 scattered
FM0200	from 0200Z
14006KT	wind 140 at 6 knots
P6SM	visibility greater than 6 statute miles
FEW025 SCT250	2,500 few, 25,000 scattered
FM0800	from 0800Z
VRB03KT	wind variable at 3 knots
5SM	visibility 5 statute miles
BR	mist
SCT012	1,200 scattered
TEMPO 1012	temporary conditions between 1000Z and 1200Z
1/2SM	visibility ½ statute mile
FG	fog
BKN001	ceiling 100 broken
FM1500	from 1500Z
17007KT	wind 170 at 7 knots
P6SM	visibility greater than 6 statute miles
SCT025=	2,500 scattered; the equal sign signifies the end of the TAF

TAF
KACK 112340Z 120024 29008KT P6SM SKC BECMG 1618 22015KT=

TAF	Aviation terminal forecast
KACK	Nantucket, Massachusetts
112340Z	prepared on the 11th at 2340Z
120024	valid period from the 12th at 0000Z until the 13th at 0000Z
29008KT	wind 290 at 8 knots
P6SM	visibility greater than 6 statute miles
SKC	sky clear
BECMG 1618	becoming between 1600Z and 1800Z
22015KT=	wind 220 at 15 knots; the equal sign signifies the end of the TAF

TAF
KMWH 200535Z 200606 NIL=

TAF	Aviation terminal forecast
KMWH	Moses Lake, Washington
200535Z	prepared on the 20th at 0535Z
200606	valid period from the 20th at 0600Z to the 21st at 0600Z
NIL=	no TAF; the equal sign signifies the end of the TAF

GUAM

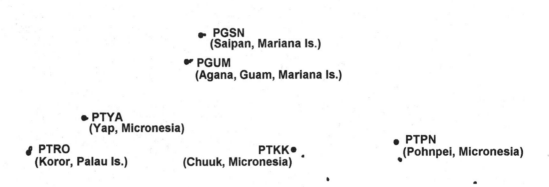

PUERTO RICO AND THE VIRGIN ISLANDS

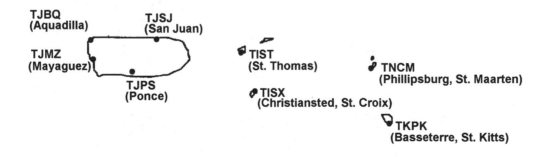

Figure 4-3. TAF Locations - Guam and Puerto Rico.

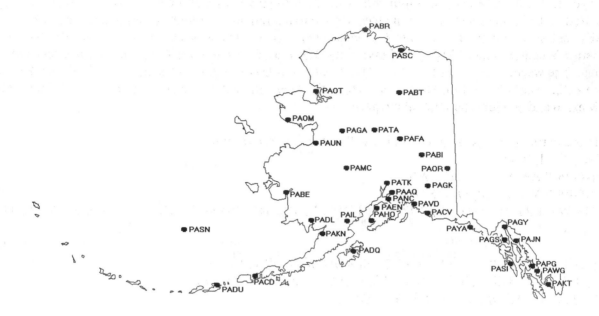

ALASKA LOCATIONS

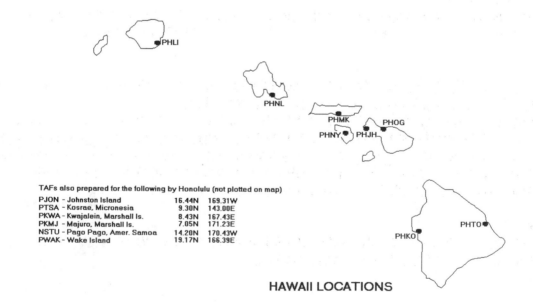

TAFs also prepared for the following by Honolulu (not plotted on map)

PJON – Johnston Island	16.44N	169.31W
PTSA – Kosrae, Micronesia	9.30N	143.00E
PKWA – Kwajalein, Marshall Is.	8.43N	167.43E
PKMJ – Majuro, Marshall Is.	7.05N	171.23E
NSTU – Pago Pago, Amer. Samoa	14.20N	170.43W
PWAK – Wake Island	19.17N	166.39E

HAWAII LOCATIONS

Figure 4-4. TAF Locations - Alaska and Hawaii.

AVIATION AREA FORECAST (FA)

An Aviation Area Forecast (FA) is a forecast of visual meteorological conditions (VMC), clouds, and general weather conditions over an area the size of several states. To understand the complete weather picture, the FA must be used in conjunction with the inflight aviation weather advisories. Together, they are used to determine forecast en route weather and to interpolate conditions at airports for which no TAFs are issued. Figure 4-5 on page 4-21 maps the FA areas. The FAs are issued 3 times a day by the Aviation Weather Center (AWC) in Kansas City, Missouri, for each of the 6 areas in the contiguous 48 states. The weather forecast office (WFO) in Honolulu issues FAs for Hawaii as shown in Figure 4-6 on page 4-22. Alaska FA information is on page 4-27. There are also two specialized FAs, one for the Gulf of Mexico and one for international airspace.

This is a partial example of an FA which will be used in this section:
DFWC FA 120945
SYNOPSIS AND VFR CLDS/WX
SYNOPSIS VALID UNTIL 130400
CLDS/WX VALID UNTIL 122200...OTLK VALID 122200-130400 OK TX AR TN LA MS AL AND
CSTL WTRS

.

SEE AIRMET SIERRA FOR IFR CONDS AND MTN OBSCN.
TS IMPLY SEV OR GTR TURB SEV ICE LLWS AND IFR CONDS.
NON MSL HGTS DENOTED BY AGL OR CIG.

.

SYNOPSIS...LOW PRES TROF 10Z OK/TX PNHDL AREA FCST MOV EWD INTO CNTRL-SWRN
  OK BY 04Z. WRMFNT 10Z CNTRL OK-SRN AR-NRN MS FCST LIFT NWD INTO NERN OK-
  NRN AR XTRM NRN MS BY 04Z.

.

S CNTRL AND SERN TX
  AGL SCT-BKN010. TOPS 030. VIS 3-5SM BR. 14-16Z BECMG AGL SCT030. 19Z AGL SCT050.
  OTLK...VFR.

.

OK
PNHDL AND NW...AGL SCT030 SCT-BKN100. TOPS FL200. 15Z AGL SCT040 SCT100. AFT 20Z
  SCT TSRA DVLPG..FEW POSS SEV. CB TOPS FL450. OTLK...VFR.
SWRN OK...CIG BKN020. TOPS 050. VIS 3-5SM BR. 14Z AGL SCT-BKN040. 18Z CIG BKN060.
  TOPS FL180. 22Z SCT TSRA DVLPG..FEW POSS SEV. CB TOPS ABV FL450. OTLK...VFR.
NERN QTR...CIG BKN020 OVC050. VIS 3-5SM NMRS TSRA..FEW POSS SEV. CB TOPS ABV
  FL450. 15Z AGL SCT030 SCT-BKN100. TOPS FL250. 18Z AGL SCT040. OTLK...VFR.
SERN QTR...AGL SCT-BKN020. TOPS 050. 18Z AGL SCT040. OTLK...VFR.

.

CSTL WTRS
LA MS AL WTRS...SCT025 SCT-BKN080. TOPS 150. ISOL -TSRA. CB TOPS FL350. OTLK...VFR.
TX WTRS...SCT CI. OCNL SCT030. OTLK...VFR. ....

The FA is comprised of four sections: a communications and product header section, a precautionary statements section, and two weather sections - a synopsis section and a visual flight rules (VFR) clouds/weather section.

COMMUNICATIONS AND PRODUCT HEADER

The communications and product header identifies the office for which the FA is issued, the date and time of issue, the product name, the valid times, and the states and/or areas covered by the FA. The following shows the communications and product header for the example FA shown on page 4-17:

DFWC FA 120945
SYNOPSIS AND VFR CLDS/WX
SYNOPSIS VALID UNTIL 130400
CLDS/WX VALID UNTIL 122200...OTLK VALID 122200-130400 OK TX AR TN LA MS AL AND
CSTL WTRS

In the first line, "DFW" indicates the area for which the FA is valid. The " **C**" indicates VFR clouds and weather while the FA indicates what type of forecast message it is. The "120945" indicates the date and time the FA was issued. The next line "SYNOPSIS AND VFR CLDS/WX" states what information is contained in this forecast message. "SYNOPSIS VALID UNTIL 130400" means the synopsis section of the FA is valid until the thirteenth at 0400Z. The "CLDS/WX VALID UNTIL 122200...OTLK VALID 122200-130400" statement indicates the forecast section is valid until the twelfth at 2200Z, while the outlook portion is valid from the twelfth at 2200Z until the thirteenth at 0400Z. "OK TX AR TN LA MS AL AND CSTL WTRS" describes the area for which this FA forecast is valid.

PRECAUTIONARY STATEMENTS

Between the communications/product header and the body of the forecast are three precautionary statements. (See example FA on page 4-17.) The first statement in the example, "SEE AIRMET SIERRA FOR IFR CONDS AND MTN OBSCN," is included to alert users that IFR conditions and/or mountain obscurations may be occurring or may be forecasted to occur in a portion of the FA area. The user shall <u>always</u> check the latest AIRMET Sierra for the FA area.

The second statement in the example, "TS IMPLY SEV OR GTR TURB SEV ICE LLWS AND IFR CONDS," is included as a reminder of the hazards existing in all thunderstorms. Thus, these thunderstorm-associated hazards are not spelled out within the body of the FA.

The purpose of the third statement in the example, "NON MSL HGTS DENOTED BY AGL OR CIG," is to alert the user that heights, for the most part, are mean sea level (MSL). All heights are in hundreds of feet. For example, "BKN030. TOPS 100. HYR TRRN OBSCD," means bases of the broken clouds are 3,000 feet MSL with tops 10,000 feet MSL. Terrain above 3,000 feet MSL will be obscured. The tops of the clouds, turbulence, icing, and freezing level heights are <u>always</u> MSL.

Heights AGL are noted in either of two ways:

1. Ceilings by definition are above ground. Therefore, the contraction "CIG" indicates above ground. For example, "CIG BKN-OVC015," means that ceilings are expected to be broken to ove rcast sky cover with bases at 1,500 feet AGL.
2. The contraction "AGL" means above ground level. Therefore, "AGL SCT020" means scattered clouds with bases 2,000 feet AGL.

Thus, if the contraction "AGL" or "CIG" is not denoted, height is automatically above MSL.

SYNOPSIS

The synopsis is a brief summary of the location and movements of fronts, pressure systems, and other circulation features for an 18-hour period. References to low ceilings and/or visibilities, strong winds, or any other phenomena the forecaster considers useful may also be included. The following synopsis is taken from the example on page 4-17.

SYNOPSIS...LOW PRES TROF 10Z OK/TX PNHDL AREA FCST MOV EWD INTO CNTRL-SWRN
   OK BY 04Z. WRMFNT 10Z CNTRL OK-SRN AR-NRN MS FCST LIFT NWD INTO NERN OK-
   NRN AR XTRM NRN MS BY 04Z.

This paragraph states that a low pressure trough at 10Z was over the Oklahoma (OK)/Texas (TX) panhandle area. The area is forecasted to move eastward into central-southwestern OK by 04Z. At 10Z a warm front was located from central OK to southern Arkansas (AR) to northern Mississippi (MS). This warm front is forecasted to lift into northeastern OK, northern AR, to extreme northern MS by 04Z.

VFR CLOUDS AND WEATHER (VFR CLDS/WX)

This section contains a 12-hour specific forecast, followed by a 6-hour categorical outlook giving a total forecast period of 18 hours, and it is usually several paragraphs in length. The breakdown may be by states or by well-known geographical areas. (See Figure 4-11.) The specific forecast section gives a general description of clouds and weather which cover an area greater than 3,000 square miles and are significant to VFR flight operations.

Surface visibility and obstructions to vision are included when the forecast visibility is 3 to 5 statute miles. Precipitation, thunderstorms, and sustained winds of 20 knots or more will always be included when forecasted. The conditional term OCNL (occasional) is used to describe clouds and visibilities that may affect VFR flights. It is used when there is a greater than 50% probability of a phenomenon occurring, but for less than ½ the forecast period.   The areal coverage terms ISOL (isolated), WDLY SCT (widely scattered), SCT or AREAS (scattered), and NMRS or WDSPRD (numerous or widespread) are used to indicate the area coverage of thunderstorms or showers. The term ISOL may also be used to describe areas of ceilings or visibilities that are expected to affect areas less than 3,000 square miles. Table 4-1 defines the areal coverage terms.

Table 4-1 Areal Coverage of Showers and Thunderstorms

Terms	Coverage
Isolated (ISOL)	Single cells (no percentage)
Widely scattered (WDLY SCT)	Less than 25% of area affected
Scattered or Areas (SCT or AREAS)	25 to 54% of area affected
Numerous or Widespread (NMRS or WDSPRD)	55% or more of area affected

Example from the FA on page 4-17:
CSTL WTRS
LA MS AL WTRS...SCT025 SCT-BKN080. TOPS 150. ISOL -TSRA. CB TOPS FL350. OTLK...VFR
TX WTRS...SCT CI. OCNL SCT030. OTLK...VFR. ....

This part of the VFR clouds/weather section is the forecast for the coastal waters of Louisiana (LA), Mississippi (MS), Alabama (AL), and Texas (TX). For the coastal waters of LA, MS, and AL, the base of the scattered layer is 2,500 feet MSL. The second layer is scattered to broken at 8,000 feet MSL with tops at 15,000 feet MSL. Also during this time, isolated (ISOL) thunderstorms with light rain showers are expected with the tops of the thunderstorms (CB) at flight level (FL) 350. FL is used only for altitudes 18,000 feet MSL and higher. The visibility is expected to be greater than 6 statute miles and winds less than 20 knots, both by omission. The weather conditions along the TX coastal waters are expected to be scattered cirrus with occasional (OCNL) scattered layers at 3,000 feet MSL.

A categorical outlook, identified by "OTLK," is included for each area breakdown. A categorical outlook of instrument flight rules (IFR) and marginal VFR (MVFR) can be due to ceilings only (CIG), restriction to visibility only (TSRA, FG, etc.), or a combination of both. In the example, the coastal areas have outlooks of VFR conditions.

The statement, "OTLK... VFR BCMG MVFR CIG F AFT 09Z," means the weather is expected to be VFR, becoming MVFR due to low ceiling, and visibilities restricted by fog after 0900Z. "WND" is included in the outlook if winds, sustained or gusty, are expected to be 20 knots or greater.

Hazardous weather (i.e., IFR, icing, and turbulence conditions) is <u>not</u> included in the FA but are included in the Inflight Aviation Weather Advisories (see page 4-23).

AMENDED AVIATION AREA FORECAST

Amendments to the FA are issued as needed. An amended FA is identified by **AMD** that is located on the first line after the date and time. The entire FA is transmitted again with the word **UPDT** after the state to indicated what sections have been amended/updated. FAs are also amended and updated by inflight aviation weather advisories (AIRMETs, SIGMETs, and Convective SIGMETs). A corrected FA is identified by **COR** and a delayed FA is identified by **RTD** which are located in the first line after the time and date.

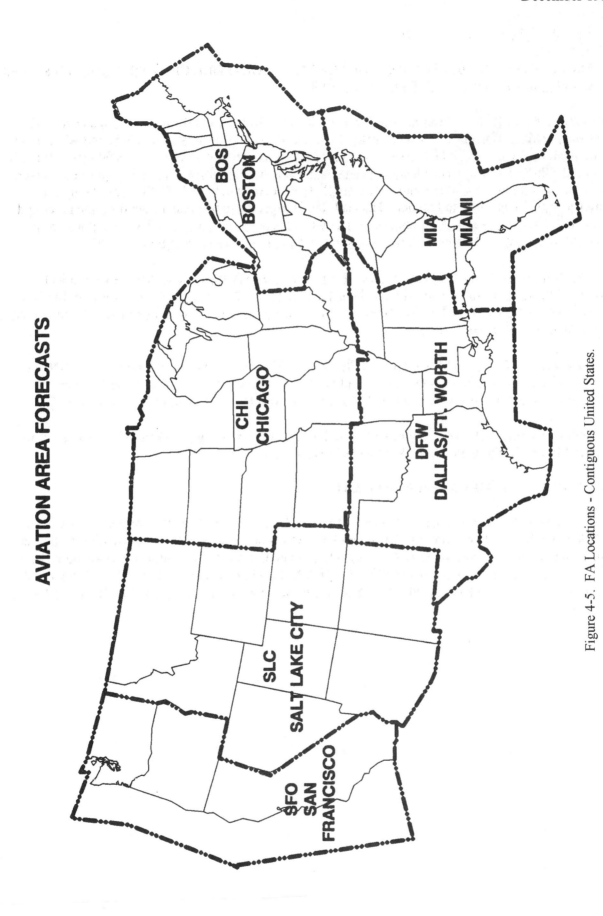

Figure 4-5. FA Locations - Contiguous United States.

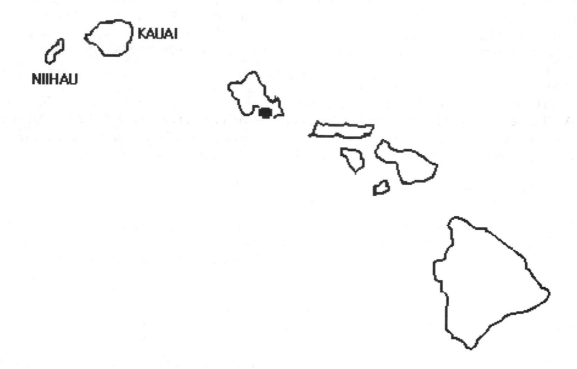

AREA FORECAST LOCATIONS - HAWAII

Figure 4-6. FA Locations - Hawaii.

INFLIGHT AVIATION WEATHER ADVISORIES

Inflight Aviation Weather Advisories are forecasts to advise en route aircraft of development of potentially hazardous weather. All inflight aviation weather advisories in the conterminous U.S. are issued by the Aviation Weather Center (AWC) in Kansas City, Missouri. The WFO in Honolulu issues advisories for the Hawaiian islands. In Alaska, the Alaska Aviation Weather Unit (AAWU) issues inflight aviation weather advisories. All heights are referenced MSL, except in the case of ceilings CIG, which indicate AGL.

There are three types of inflight aviation weather advisories - the Significant Meteorological Information (SIGMET), the Airman's Meteorological Information (AIRMET), and Convective SIGMET. All of these advisories use the same location identifiers (either VORs, airports, or well-known geographic areas) to describe the hazardous weather areas (see Figures 4-11 and 4-12 on pages 4-45 and 4-46).

SIGMET (WS)/AIRMET (WA)

SIGMETs/AIRMETs are issued corresponding to the FA areas (see Figures 4-5 and 4-6). The maximum forecast period is 4 hours for SIGMETs and 6 hours for AIRMETs. Both advisories are considered "widespread" because they must be either affecting or be forecasted to affect an area of at least 3,000 square miles at any one time. However, if the total area to be affected during the forecast period is very large, it could be that in actuality only a small portion of this total area would be affected at any one time.

SIGMET (WS)

A SIGMET advises of nonconvective weather that is potentially hazardous to all aircraft. SIGMETs are unscheduled products that are valid for 4 hours. However, conditions that are associated with hurricanes are valid for 6 hours. Unscheduled updates and corrections are issued as necessary. In the conterminous U.S., SIGMETs are issued when the following phenomena occur or are expected to occur:

1. Severe icing not associated with thunderstorms
2. Severe or extreme turbulence or clear air turbulence (CAT) not associated with thunderstorms
3. Dust storms or sandstorms lowering surface or inflight visibilities to below 3 miles
4. Volcanic ash

In Alaska and Hawaii, SIGMETs are also issued for:

1. Tornadoes
2. Lines of thunderstorms
3. Embedded thunderstorms
4. Hail greater than or equal to ¾ inch in diameter

SIGMETs are identified by an alphabetic designator from November through Yankee excluding Sierra and Tango. (Sierra, Tango, and Zulu are reserved for AIRMETs.) The first issuance of a SIGMET will be labeled as UWS (Urgent Weather SIGMET). Subsequent issuances are at the forecaster's discretion. Issuance for the same phenomenon will be sequentially numbered, using the original designator until the phenomenon ends. For example, the first issuance in the Chicago (CHI) FA area for phenomenon moving from the Salt Lake City (SLC) FA area will be SIGMET Papa 3, if the previous two issuances, Papa 1 and Papa 2, had been in the SLC FA area. Note that no two different phenomena across the country can have the same alphabetic designator at the same time.

Example of a SIGMET:
BOSR WS 050600
SIGMET ROMEO 2 VALID UNTIL 051000
ME NH VT
FROM CAR TO YSJ TO CON TO MPV TO CAR
MOD TO OCNL SEV TURB BLW 080 EXP DUE TO STG NWLY FLOW. CONDS CONTG BYD
1000Z.

International SIGMET

Some NWS offices have been designated by the ICAO as Meteorological Watch Offices (MWOs).
These offices are responsible for issuing International SIGMETs for designated areas that include
Alaska, Hawaii, portions of the Atlantic and Pacific Oceans, and the Gulf of Mexico. The offices which
issue International SIGMETs are the Alaskan Aviation Weather Unit in Anchorage, Alaska (AK);the
Tropical Prediction Center in Miami, Florida (FL); the WFO in Honolulu, Hawaii (HI); the Aviation
Weather Center in Kansas City, MO; and the WFO on Guam Island in the Pacific Ocean. These
SIGMETs are considered "widespread" because they must be either affecting or be forecasted to affect
an area of at least 3,000 square miles at any one time. The International SIGMET is issued for 12 hours
for volcanic ash events, 6 hours for hurricanes and tropical storms, and 4 hours for all other events. Like
the domestic SIGMETs, International SIGMETs are also identified by an alphabetic designator from
Alpha through Mike and are numbered sequentially until that weather phenomenon ends. The criteria for
an International SIGMET are:

1. Thunderstorms occurring in lines, embedded in clouds, or in large areas producing tornadoes or large
   hail
2. Tropical cyclones
3. Severe icing
4. Severe or extreme turbulence
5. Dust storms and sandstorms lowering visibilities to less than 3 miles
6. Volcanic ash

Example of an International SIGMET:
ZCZC MIASIGA1L
TTAA00 KNHC 121600

KZNY SIGMET LIMA 5 VALID 121600/122000 UTC KNHC-

ACT TS OBS BY SATELLITE WI AREA BOUNDED BY 30N69W 31N64.6W 26.4N66.4W
27.5N69.4W 30N69W. CB TOPS TO FL480. MOV ENE 15 KT. INTSF.

AIRMET (WA)

AIRMETs (WAs) are advisories of significant weather phenomena but describe conditions at intensities
lower than those which require the issuance of SIGMETs. AIRMETs are intended for dissemination to
all pilots in the preflight and en route phase of flight to enhance safety. AIRMET Bulletins are issued on
a scheduled basis every 6 hours beginning at 0145 UTC during Central Daylight Time and at 0245 UTC
during Central Standard Time. Unscheduled updates and corrections are issued as necessary. Each
AIRMET Bulletin contains any current AIRMETs in effect and an outlook for conditions expected after
the AIRMET valid period. AIRMETs contain details about IFR, extensive mountain obscuration,
turbulence, strong surface winds, icing, and freezing levels.

There are three AIRMETs - Sierra, Tango, and Zulu. AIRMET Sierra describes IFR conditions and/or extensive mountain obscurations. AIRMET Tango describes moderate turbulence, sustained surface winds of 30 knots or greater, and/or nonconvective low-level wind shear. AIRMET Zulu describes moderate icing and provides freezing level heights. After the first issuance each day, scheduled or unscheduled bulletins are numbered sequentially for easier identification.

Example of AIRMET Sierra issued for the Chicago FA area:
CHIS WA 121345
AIRMET SIERRA UPDT 3 FOR IFR AND MTN OBSCN VALID UNTIL 122000 .
AIRMET IFR...SD NE MN IA MO WI LM MI IL IN KY
FROM 70NW RAP TO 50W RWF TO 50W MSN TO GRB TO MBS TO FWA TO CVG TO HNN TO TRI TO ARG TO 40SSW BRL TO OMA TO BFF TO 70NW RAP
OCNL CIG BLW 010/VIS BLW 3SM FG/BR. CONDS ENDG 15Z-17Z.

AIRMET MTN OBSCN...KY TN
FROM HNN TO TRI TO CHA TO LOZ TO HNN
MTNS OCNL OBSC CLDS/PCPN/BR. CONDS ENDG TN PTN AREA 18Z- 20Z.CONTG KY BYD 20Z..ENDG 02Z.

....

Example of AIRMET Tango issued for the Salt Lake City FA area:
SLCT WA 121345
AIRMET TANGO UPDT 2 FOR TURB VALID UNTIL 122000 .
AIRMET TURB...NV UT CO AZ NM
FROM LKV TO CHE TO ELP TO 60S TUS TO YUM TO EED TO RNO TO LKV
OCNL MOD TURB BLW FL180 DUE TO MOD SWLY/WLY WNDS. CONDS CONTG BYD 20Z THRU 02Z.

AIRMET TURB...NV WA OR CA CSTL WTRS
FROM BLI TO REO TO BTY TO DAG TO SBA TO 120W FOT TO 120W TOU TO BLI
OCNL MOD TURB BTWN FL180 AND FL400 DUE TO WNDSHR ASSOCD WITH JTSTR. CONDS CONTG BYD 20Z THRU 02Z.

....

Example of AIRMET Zulu issued for the San Francisco FA area:
SFOZ WA 121345
AIRMET ZULU UPDT 2 FOR ICE AND FRZLVL VALID UNTIL 122000 .
AIRMET ICE...WA OR ID MT NV UT
FROM YQL TO SLC TO WMC TO LKV TO PDT TO YDC TO YQL
LGT OCNL MOD RIME/MXD ICGICIP BTWN FRZLVL AND FL220.FRZLVL 080-120. CONDS CONTG BYD 20Z THRU 02Z.

AIRMET ICE...WA OR
FROM YDC TO PDT TO LKV TO 80W MFR TO ONP TO TOU TO YDC
LGT OCNL MOD RIME/MXD ICGICIP BTWN FRZLVL AND FL180.FRZLVL 060-080. CONDS CONTG BYD 20Z THRU 02Z.

FRZLVL...WA...060 CSTLN SLPG 100 XTRM E.
OR...060-070 CASCDS WWD. 070-095 RMNDR.
NRN CA...060-100 N OF A 30N FOT-40N RNO LN SLPG 100-110 RMNDR.

CONVECTIVE SIGMET (WST)

Convective SIGMETs are issued in the conterminous U.S. for any of the following:

1. Severe thunderstorm due to:
    a. surface winds greater than or equal to 50 knots
    b. hail at the surface greater than or equal to ¾ inches in diameter
    c. tornadoes
2. Embedded thunderstorms
3. A line of thunderstorms
4. Thunderstorms producing precipitation greater than or equal to heavy precipitation affecting 40% or more of an area at least 3,000 square miles

Any convective SIGMET implies severe or greater turbulence, severe icing, and low-level wind shear. A convective SIGMET may be issued for any convective situation that the forecaster feels is hazardous to all categories of aircraft.

Convective SIGMET bulletins are issued for the western (W), central (C), and eastern (E) United States. (Convective SIGMETs are not issued for Alaska or Hawaii.) The areas are separated at 87 and 107 degrees west longitude with sufficient overlap to cover most cases when the phenomenon crosses the boundaries. Bulletins are issued hourly at H+55. Special bulletins are issued at any time as required and updated at H+55. If no criteria meeting convective SIGMET requirements are observed or forecasted, the message "CONVECTIVE SIGMET… NONE" will be issued for each area at H+55. Individual convective SIGMETs for each area (W, C, E) are numbered sequentially from number one each day, beginning at 00Z. A convective SIGMET for a continuing phenomenon will be reissued every hour at H+55 with a new number. The text of the bulletin consists of either an observation and a forecast or just a forecast. The forecast is valid for up to 2 hours.

Example of a convective SIGMET:
MKCC WST 251655
CONVECTIVE SIGMET 54C
VALID UNTIL 1855Z
WI IL
FROM 30E MSN-40ESE DBQ
DMSHG LINE TS 15 NM WIDE MOV FROM 30025KT. TOPS TO FL450. WIND GUSTS TO 50 KT POSS.

CONVECTIVE SIGMET 55C
VALID UNTIL 1855Z
WI IA
FROM 30NNW MSN-30SSE MCW
DVLPG LINE TS 10 NM WIDE MOV FROM 30015KT. TOPS TO FL300.

CONVECTIVE SIGMET 56C
VALID UNTIL 1855Z
MT ND SD MN IA MI
LINE TS 15 NM WIDE MOV FROM 27020KT. TOPS TO FL380.

OUTLOOK VALID 151855-252255
FROM 60NW ISN-INL-TVC-SBN-BRL-FSD-BIL-60NW ISN

IR STLT IMGRY SHOWS CNVTV CLD TOP TEMPS OVER SRN WI HAVE BEEN WARMING STEADILY INDCG A WKNG TREND. THIS ALSO REFLECTED BY LTST RADAR AND LTNG DATA. WKNG TREND OF PRESENT LN MAY CONT… HWVR NEW DVLPMT IS PSBL ALG OUTFLOW BDRY AND/OR OVR NE IA/SW WI BHD CURRENT ACT.

A SCND TS IS CONTG TO MOV EWD THRU ERN MT WITH NEW DVLPMT OCRG OVR CNTRL ND. MT ACT IS MOVG TWD MORE FVRBL AMS OVR THE WRN DAKS WHERE DWPTS ARE IN THE UPR 60S WITH LIFTED INDEX VALUES TO MS 6. TS EXPD TO INCR IN COVERAGE AND INTSTY DURG AFTN HRS.

WST ISSUANCES EXPD TO BE RQRD THRUT AFTN HRS WITH INCRG PTNTL FOR STGR CELLS TO CONTAIN LRG HAIL AND PSBLY DMGG SFC WNDS.

ALASKA, GULF OF MEXICO, AND INTERNATIONAL AREA FORECASTS (FAs)

ALASKA AREA FORECAST (FA)

The Alaska Aviation Weather Unit in Anchorage, Alaska, produces the FA for the entire state of Alaska. The Alaska FA combines the FA, SIGMETs, and AIRMETs into one product. Each FA contains a regional synopsis, 12-hour geographic specific forecasts, and an 18-hour outlook for each geographic area. Forecast weather elements are sky condition, cloud height, mountain obscuration, visibility, weather and/or obstructions to visibility, strong surface winds (direction and speed), icing, freezing level, and mountain pass conditions. Hazards and flight precautions, including AIRMETs and SIGMETs, may be found in their respective geographic area. AIRMETs and SIGMETs are also issued as separate products.

Partial example of Alaska FA:
JNUH FA 191445

.

EASTERN GULF COAST AND SE AK…

.

AIRMET VALID UNTIL 230300
TS IMPLY POSSIBLE SEV OR GREATER TURB SEV ICE LLWS AND IFR CONDS.
NON MSL HEIGHTS NOTED BY AGL OR CIG

.

SYNOPSIS… VALID UNTIL 231500
990 MB LOW VCY PACV DRFTG E AND WKN. CDFNT S FM LOW BCMG STNR AND WK ICY BAY SWD BY 15Z. E PACIFIC LOW S 50N MOV N TO 975 MB CNTR 50 SM W PASI AT 15Z WI OCFNT SWD.

.

LYNN CANAL AND GLACIER BAY JB… VALID UNTIL 230900
… CLOUDS/WX…
… AIRMET MT OBSC… TEMPO MT OBSC INCLDS. NC…
SCT030 SCT-BKN050 BKN100 TOP 160. TEMPO HI LYRS TOP FL250. TEMPO BKN030 ISOL -RA. SFC WND S 15 KT G25 KT LYNN CANAL.
OTLK VALID 230900-240300… VFR RA. 18Z MVFR CIG RA.
PASSES… WHITE AND CHILKOOT… MVFR CIG RASN.
… TURB…
LYNN CANAL… ISOL MOD TURB BLW 060. ELSW.NIL SIG.
… ICE AND FZLVL…
TEMPO LGT RIME ICEIC 050-120. FZLVL 030.

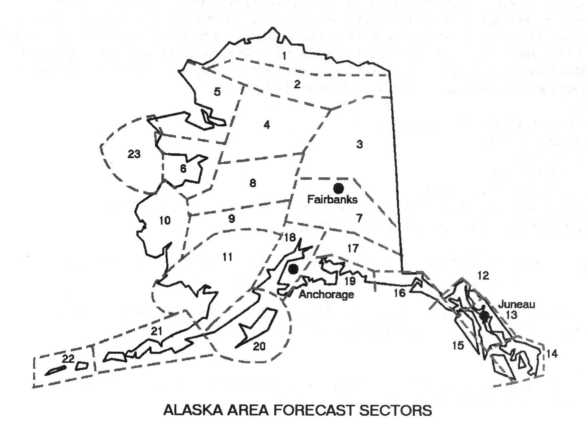

ALASKA AREA FORECAST SECTORS

Figure 4-7. Alaska Area Forecast Sectors.

GULF OF MEXICO AREA FORECAST

A specialized FA for the Gulf of Mexico is issued by the Tropical Prediction Center in Miami, Florida. The product combines the FA, inflight aviation weather advisories, and marine precautions. This product is intended to support both offshore heliport and general aviation operations. The Gulf of Mexico FA focuses on an area which includes the coastal plains and coastal waters from Apalachicola, Florida, to Brownsville, Texas, and the offshore waters of the Gulf of Mexico, in an area west of 85W longitude and north of 27N latitude. Each section of the FA describes the weather conditions expecting to impact the area and will include the descriptor <u>none</u> if no significant weather is forecast to occur. Amendments to this FA are issued the same as amendments to the domestic FAs.

Partial example of Gulf of Mexico FA:
FAGX01 KNHC 151030
151100Z-152300Z
OTLK... 152300Z-161100Z
AMDT NOT AVBL 0200Z-1100Z
TROPICAL ANALYSIS AND FORECAST BRANCH
TROPICAL PREDICTION CENTER MIAMI FLORIDA

GLFMEX N OF 27N W OF 85W... CSTL PLAINS CSTL WTRS AQQ-BRO. HGTS MSL UNLESS NOTED.

TS IMPLY POSS SEV OR GTR TURB... SEV ICE... LOW LVL WS AND STG SFC WND... HIGH WAVES... CIG BLW 010... AND VIS BLW 3SM.

01 SYNS...
WK SFC TROUGH FM 31N84W TO 26N88W AT 11Z DRIFTING E THROUGH 23Z. WK HIGH PRES ACRS RMNDER GLFMEX THRU FCST AND OTLK PD.
...
02 FLT PRCTNS...
NONE.
...

03 MARINE PRCTNS...
NONE.
...
04 SGFNT CLD/WX...
CSTL PLAINS CSTL WTRS BRO-LCH AND OFSHR WTRS W OF 94W... FEW040. OTLK... VFR.
...
CSTL PLAINS LCH-AQQ...
FEW015. OCNL VIS 3-5SM BR. AFT 14Z SCT100. AFT 19Z SCT/BKN020-030 BKN/SCT070-090. WIDELY SCT TSRA/ISOL +TSRA.
...
05 ICE AND FZ LEVEL BLW 120...
NONE. FZ LEVEL ABV 120.
...
06 TURB BLW 120...
NONE.
...

07 WND BLW 120...
CSTL PLAINS CSTL WTRS LCH-GPT AND OFSHR WTRS 94W-89W... SFC-120 NE-E 10-15 KT.
OTLK... NOSIG.
...
08 WAVES...
CSTL WTRS BRO-AQQ... 1-2 FT. OTLK... NOSIG.
NNNN

INTERNATIONAL AREA FORECASTS

FAs from the surface to 25,000 feet are also prepared in international format for areas in the Atlantic Ocean, Caribbean Sea, and the Gulf of Mexico. Moreover, significant weather forecasts for 25,000 feet to 60,000 feet are prepared in chart form and in international text format for the Northern and Western hemispheres.

Example of an International FA from the surface to FL250:
FANT2 KWBC 091600
091800Z TO 100600Z

ATLANTIC OCEAN WEST OF A LINE FROM 40N 67W TO 32N 63W. SFC TO FL250.

SYNOPSIS.
RIDGE OVER AREA MOVING TO EAST. FRONTAL SYSTEM MOVING OFF COAST BY 06Z.

SIGNIFICANT CLDS/WX.
N OF 34N AND W OF 71W... PATCHES OVC005/015 TOP 030/040 OTHERWISE
BKN/OVC015/025 BKN/OVC200/240. BY 06Z INCREASING IMC IN SHRA/TS SPREADING
ACROSS AREA FROM WEST. TS TOPS ABOVE 240.

S OF 34N AND W OF 75W... SCT/BKN 015/250. BY 06Z INCREASING IMC IN SHRA/TS
SPREADING ACROSS AREA FROM WEST. TS TOPS ABOVE 240.

ELSEWHERE... CLR OCNL SCT015/025. BY 06Z INCREASING BKN080/100.

ICE.
FZ LVL 080/090 N SLOPING TO 120/130 S. MOD IN SHRA. SEV IN TS.

TURB.
MOD IN SHRA. SEV IN TS.

OUTLOOK.
100600Z TO 101800Z
FRONT CONTINUING SLOWLY EWD. INCREASING IMC IN SHRA/TS SPREADNG E OVER
AREA. SHRA/TS ENDING SW PORTION AFTER FRONTAL PASSAGE.

Example of international significant weather forecast for FL250 to FL600:
FAPA1 KWBC 141610
SIG WX PROG FL250-FL600 VALID 150600Z
ISOL EMBD CB TOPS 400 NE OF 11N173W 14N166W 11N164W 01N174W
ISOL EMBD CB TOPS 400 07N158W 08N137W 11N137W 12N158W 07N158W
ISOL EMBD CB TOPS 400 19N157W 32N143W 22N162W 15N162W
MDT OR GRTR TURB AND ICG VCNTY ALL CBS
MDT TURB 310-410 19N145W 25N144W 19N163W 15N162W 19N145W

The groups of numbers and letters are the boundary points of the areas in latitude and longitude. For example, "11N173W" is latitude 11 degrees north and longitude 173 degrees west.

TRANSCRIBED WEATHER BROADCAST (TWEB) TEXT PRODUCTS

NWS offices prepare transcribed weather broadcast (TWEB) text products for the contiguous U.S., including synopsis and forecast for more than 200 routes and local vicinities. (See Figure 4-8.) (Not all NWS forecast offices issue all three of these products.) These products may be used in the Telephone Information Briefing Service (TIBS), Pilot's Automatic Telephone Weather Answering Service (PATWAS), Low/Medium Frequency (L/MF) and VHF omni-directional radio range (VOR) facilities as described in Section 1. TWEB products are valid for 12 hours and are issued 4 times a day at 0200Z, 0800Z, 1400Z, and 2000Z. A TWEB route forecast or vicinity forecast will not be issued if the TAF for that airport has not been issued. A NIL TWEB will be issued instead.

A TWEB route forecast is for a 50-nautical-mile wide corridor along a line connecting the end points of the route. A TWEB local vicinity forecast covers an area with a radius of 50 nautical miles. The route and vicinity forecasts describe specific information on sustained surface winds (25 knots or greater), visibility, weather and obscuration to vision, sky conditions (coverage and ceiling/cloud heights), mountain obscurement, and nonconvective low-level wind shear. If visibility of 6SM or less is forecast, obstructions to vision and/or weather will be included. Thunderstorms and volcanic ash will always be included regardless of the visibility. Cloud bases can be described either in MSL or AGL (CIGS or BASES). It depends on which statement is used: "ALL HGTS MSL XCP CIGS." or "ALL HGTS AGL XCP TOPS." Use of "AGL," "CIGS," and "BASES" should be limited to cloud bases within 4,000 feet of the ground. Cloud tops, referenced to MSL, should also be forecasted following the sky cover when expected to be below 15,000 MSL using the sky cover contractions FEW, SCT, or BKN. Nonconvective low-level wind shear will be included when the TAF for the airport involved has issued a nonconvective low-level wind shear forecast. Expected areas of icing and turbulence will not be included.

Example of TWEB route forecast:
249 TWEB 251402 KISN-KMOT-KGFK. ALL HGTS AGL XCP TOPS. KISN-50NM E KISN TIL 00Z P6SM SKC... AFT 00Z P6SM SCT050 LCL P6SM -TSRA BKN050. 50NM E KISN-KDVL TIL 20Z P6SM SCT070... AFT 20Z P6SM SCT070 LCL SFC WNDS VRB35G45KT 3-5SM TSRA CIGS OVC030-040. KDVL-KGFK TIL 16Z P6SM SCT-BKN020 AREAS 3-5SM BR... AFT 16Z P6SM SCT040.

Explanation of route forecast:
249 - route number
TWEB - TWEB route forecast
25 - twenty-fifth day of the month
1402 - valid 14Z on the twenty-fifth to 02Z on the twenty-sixth (12 hours)
KISN-KMOT-KGFK - route: Williston, North Dakota (ND), to Minot, ND, to Grand Forks, ND
Remainder of the message explained: All heights AGL except cloud tops. KISN-50NM E KISN until 00Z, visibility greater than 6SM with clear skies. After 00Z, visibility greater than 6SM with scattered clouds at 5,000 feet AGL. Local areas of visibility greater than 6SM, thunderstorm with light rain showers, and broken clouds at 5,000 feet AGL. 50 NM E KISN-KDVL (Devil's Lake, ND) until 20Z, visibility greater than 6SM, scattered clouds at 7,000 feet AGL. After 20Z, visibility greater than 6SM, scattered clouds at 7,000 feet AGL, local surface winds variable at 35 gusting to 45 knots, visibility 3-5SM, thunderstorm with moderate rain showers, overcast ceilings 3,000-4,000 feet AGL. KDVL-KGFK until 16Z, visibility greater than 6SM, scattered to broken clouds at 2,000 feet AGL, areas of visibility 3-5SM with mist. After 16Z, visibility greater than 6SM, scattered clouds at 4,000 feet AGL.

An example of TWEB vicinity forecast:
431 TWEB 021402 LAX BASIN. ALL HGTS MSL XCP CIGS. TIL 18Z P6SM XCP 3SM BR VLYS BKN020... 18Z-22Z P6SM SCT020 SCT-BKN100... AFT 22Z P6SM SKC.

Explanation of vicinity forecast:
431 - TWEB vicinity number
TWEB - TWEB forecast
02 - second day of the month
1402 - valid 14Z on the second to 02Z on the third (12 hours)
LAX BASIN - The weather conditions in the Los Angeles basin until 18Z, visibility greater than 6SM except 3SM due to mist in the valleys and broken clouds at 2,000 feet MSL. Between 18Z and 22Z, visibility greater than 6SM and scattered clouds at 2,000 feet AGL; also scattered to broken clouds at 10,000 feet MSL. After 22Z, visibility greater than 6SM and sky clear.

A TWEB synopsis forecast is a brief description of the weather systems affecting the route during the forecast valid period. The synopsis describes movement of pressure systems, movement of fronts, upper air disturbances, or air flow.

An example of a TWEB synopsis:
BIS SYNS 250820. LO PRES TROF MVG ACRS ND TDA AND TNGT. HI PRES MVG SEWD FM CANADA INTO NWRN ND BY TNGT AND OVR MST OF ND BY WED MRNG.

Explanation of synopsis:
BIS - Bismarck, ND, WFO issuing the synopsis and route forecast
SYNS - Synopsis for the area covered by the route forecast
25 - twenty-fifth day of the month
0820 - Valid from 08Z on the twenty-fifth to 20Z on the twenty-fifth (12 hours)
The remainder of message explained: Low pressure trough moving across North Dakota today and tonight. High pressure moving southeastward from Canada into northwestern North Dakota by tonight and over most of North Dakota by Wednesday morning.

An example of another TWEB synopsis:
CYS SYNS 101402 STG UPSLP WNDS OVR WY TIL 01Z WITH WDSPRD IFR CONDS IN LGT SN AND BLOWING SN. CONDS WL IPV FM N TO S ACRS WY AFT 01Z WITH DCRG CLDS.

Explanation of synopsis:
CYS - Cheyenne, WY, WFO issuing the synopsis and route forecast
SYNS - Synopsis for the area covered by the route forecast
10 - tenth day of the month
1402 - Valid from 14Z on the tenth to 02Z on the eleventh (12 hours)
The remainder of the message explained: Strong upslope winds over Wyoming until 01Z with widespread IFR conditions in light snow and blowing snow. Conditions will improve from north to south across Wyoming after 01Z with decreasing clouds.

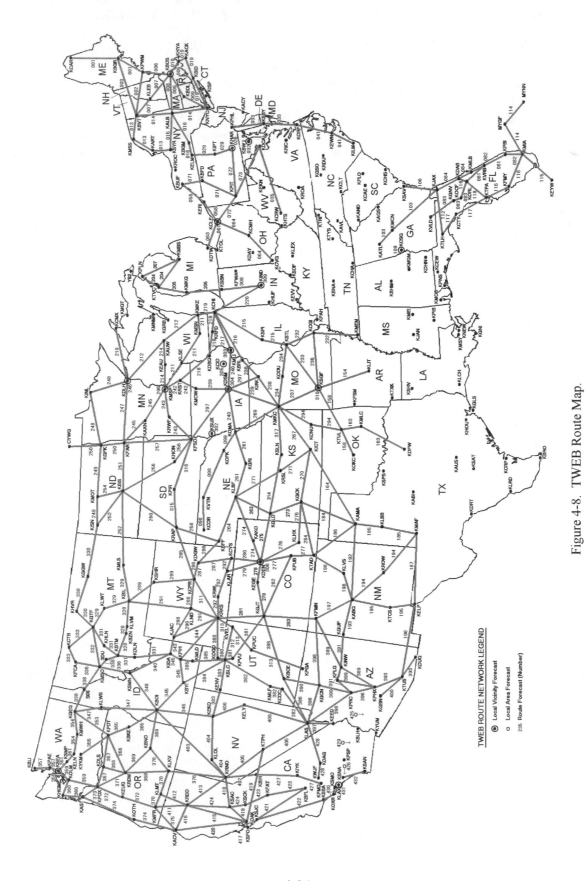

Figure 4-8. TWEB Route Map.

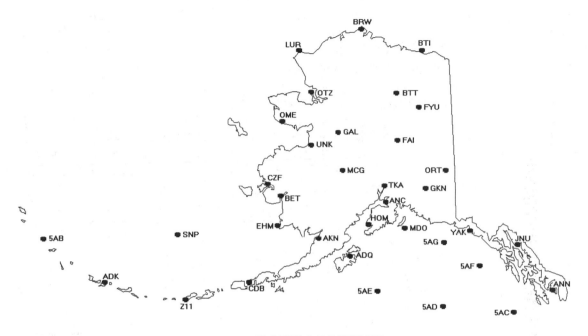

ALASKA LOCATIONS

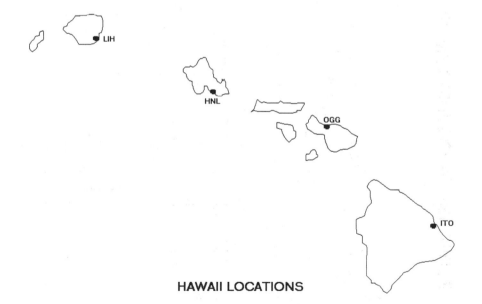

HAWAII LOCATIONS

Figure 4-10. FD Locations for Alaska and Hawaii.

CENTER WEATHER SERVICE UNIT (CWSU) PRODUCTS

Center Weather Service Unit (CWSU) products are issued by the CWSU meteorologists located in the Air Route Traffic Control Centers (ARTCCs). Coordination between the CWSU meteorologist and other NWS facilities is extremely important since both can address the same event. If time permits, coordination should take place before the CWSU meteorologist issues a product.

METEOROLOGICAL IMPACT STATEMENT (MIS)

A Meteorological Impact Statement (MIS) is an unscheduled flow control and flight operations planning forecast. The MIS can be valid between 2 to 12 hours after issuance. This enables the impact of expected weather conditions to be included in air traffic control decisions in the near future. The MIS will be issued when the following three conditions are met:

1. If any one of the following conditions occur, are forecasted to occur, and if previously forecasted, are no longer expected to occur:
    a. convective SIGMET criteria
    b. moderate or greater icing and/or turbulence
    c. heavy or freezing precipitation
    d. low IFR conditions
    e. surface winds/gusts 30 knots or greater
    f. low-level wind shear within 2,000 feet of the surface
    g. volcanic ash, dust or sandstorm
2. If the impact occurs on air traffic flow within the ARTCC area of responsibility
3. If the forecast lead time (the time between issuance and onset of a phenomenon), in the forecaster's judgment, is sufficient to make issuance of a Center Weather Advisory (CWA) unnecessary

Example of a MIS:
ZOA MIS 01 VALID 041415-041900
… FOR ATC PLANNING PURPOSES ONLY…
FOR SFO BAY AREA
DNS BR/FG WITH CIG BLO 005 AND VIS OCNL BLO 1SM TIL 19Z.

This MIS from the Fremont, California (CA), ARTCC is the first issuance of the day. It was issued at 1415Z on the fourth and is valid until 1900Z on the fourth. This forecast is for the San Francisco Bay area. The forecast is of dense fog/mist with ceilings below 500 feet and visibilities occasionally below 1SM until 19Z.

Example:
ZOA MIS 02 VALID 041650
… FOR ATC PLANNING PURPOSES ONLY…
FOR SFO BAY AREA
CANCEL ZOA MIS 01. DNS BR/FG CONDS HAVE IPVD ERYR THAN FCST.

This MIS is from the Fremont, CA, ARTCC and cancels the previously issued MIS. Specifically it states dense fog/mist conditions have improved earlier than forecasted.

Example:
ZID MIS 03 VALID 041200-042330
... FOR ATC PLANNING PURPOSES ONLY...
FROM IND TO CMH TO LOZ TO EVV TO IND
FRQ MOD TURBC FL310-390 DUE TO JTSTR... CONDS DMSHG IN INTSTY AFT 21Z.

This MIS from the Indianapolis, Indiana (IN), ARTCC was issued at 1200Z on the fourth and valid until the fourth at 2330Z. This forecast describes an area from Indianapolis, IN, to Columbus, Ohio (OH), to London, Kentucky (KY), to Evansville, IN, and back to Indianapolis, IN. The MIS describes frequent moderate turbulence between flight levels 310-390 caused by the jet stream. Conditions will diminish in intensity after 21Z.

CENTER WEATHER ADVISORY (CWA)

A Center Weather Advisory (CWA) is an aviation warning for use by air crews to anticipate and avoid adverse weather conditions in the en route and terminal environments. The CWA is not a flight planning product; instead it reflects current conditions expected at the time of issuance and/or is a short-range forecast for conditions expected to begin within 2 hours of issuance. CWAs are valid for a maximum of 2 hours. If conditions are expected to continue beyond the 2-hour valid period, a statement will be included in the CWA.

A CWA may be issued for the following three situations:

1. As a supplement to an existing inflight aviation weather advisory for the purpose of improving or updating the definition of the phenomenon in terms of location, movement, extent, or intensity relevant to the ARTCC area of responsibility. This is important for the following reason. A SIGMET for severe turbulence was issued by AWC, and the outline covered the entire ARTCC area for the total 4-hour valid time period. However, the forecaster may issue a CWA covering only a relatively small portion of the ARTCC area at any one time during the 4-hour period.
2. When an inflight aviation weather advisory has not yet been issued but conditions meet the criteria based on current pilot reports and the information must be disseminated sooner than the AWC can issue the inflight aviation weather advisory. In this case of an impending SIGMET, the CWA will be issued as urgent (UCWA) to allow the fastest possible dissemination.
3. When inflight aviation weather advisory criteria are not met but conditions are or will shortly be adversely affecting the safe flow of air traffic within the ARTCC area of responsibility.

Example of a CWA:
ZME1 CWA 081300
ZME CWA 101 VALID UNTIL 081500
FROM MEM TO JAN TO LIT TO MEM
AREA SCT VIP 5-6 (INTENSE/EXTREME) TS MOV FROM 26025KT. TOPS TO FL450.

This CWA was issued by the Memphis, Tennessee (TN), ARTCC. The 1 after the ZME in the first line denotes this CWA has been issued for the first weather phenomenon to occur for the day. It was written on the eighth at 1300Z. The 101 in the second line denotes the phenomenon number again (1) and the issuance number (01) for this phenomenon. The CWA is until the eighth at 1500Z. The area is bounded from Memphis, TN, to Jackson, MS, to Little Rock, AR, and back to Memphis, TN. Within the CWA is an area with scattered VIP 5-6 (intense/extreme) thunderstorms moving from 260 degrees at 25 knots. Tops of the thunderstorms are at FL450.

HURRICANE ADVISORY (WH)

When a hurricane threatens a coastline, but is located at least 300NM offshore, a Hurricane Advisory (WH) is issued to alert aviation interests. The advisory gives the location of the storm center, its expected movement, and the maximum winds in and near the storm center. It does not contain details of associated weather, as specific ceilings, visibilities, weather, and hazards that are found in the FAs, TAFs, and inflight aviation weather advisories.

Example of a WH:
ZCZC MIATCPAT4
TTAA00 KNHC 190841
BULLETIN
HURRICANE DANNY ADVISORY NUMBER 13
NATIONAL WEATHER SERVICE MIAMI FL
4 AM CDT SAT JUL 19 1997

… DANNY STILL MOVING LITTLE… ANY NORTHWARD DRIFT WOULD BRING THE CENTER ONSHORE…

HURRICANE WARNINGS ARE IN EFFECT FROM GULFPORT MISSISSIPPI TO APALACHICOLA FLORIDA. SMALL CRAFT SOUTHWEST OF GULFPORT SHOULD REMAIN IN PORT UNTIL WINDS AND SEAS SUBSIDE.

AT 4 AM CDT… 0900Z… THE CENTER OF HURRICANE DANNY WAS LOCATED BY NATIONAL WEATHER SERVICE RADAR AND RECONNAISSANCE AIRCRAFT NEAR LATITUDE 30.2 NORTH… LONGITUDE 88.0 WEST… VERY NEAR THE COAST ABOUT 25 MILES SOUTH-SOUTHEAST OF MOBILE ALABAMA.

DANNY HAS MOVED LITTLE DURING THE PAST FEW HOURS. WHILE SOME ERRATIC MOTION CAN BE EXPECTED DURING THE NEXT FEW HOURS… A GRADUAL TURN TOWARD THE NORTHEAST IS EXPECTED TODAY. ON THIS COURSE… THE CENTER IS EXPECTED TO MAKE LANDFALL IN THE WARNING AREA TODAY. HOWEVER ANY DEVIAITON TO THE NORTH OR THE TRACK WOULD BRING THE CENTER ONSHORE WITHIN THE WARNING AREA AT ANYTIME.
MAXIMUM SUSTAINED WINDS ARE NEAR 75 MPH WITH HIGHER GUSTS. SOME STRENGTHENING IS STILL POSSIBLE PRIOR TO LANDFALL. DAUPHIN ISLAND RECENTLY REPORTED GUSTS TO 66 MPH AND THE PRESSURE DROPPED TO 989MB… 29.20 INCHES.

DANNY HAS A RELATIVELY SMALL WIND FIELD. HURRICANE FORCE WINDS EXTEND OUTWARD UP TO 25 MPH FROM THE CENTER AND TROPICAL STORM FORCE WINDS EXTEND OUTWARD UP TO 70 MILES.

LATEST MINIMUM CENTRAL PRESSURE REPORTED BY A RECONNAISSANCE AIRCRAFT WAS 986 MB… 29.11 INCHES.

RADAR SHOWS RAIN BANDS AFFECTING THE AREA FROM SOUTHERN MISSISSIPPI TO THE FLORIDA PANHANDLE. TOTALS OF 10 TO 20 INCHES… LOCALLY HIGHER… COULD OCCUR NEAR THE TRACK OF DANNY DURING THE NEXT FEW DAYS.

STORM SURGE FLOODING OF 4 TO 5 FEET ABOVE NORMAL TIDES IS POSSIBLE ALONG THE GULF COAST EAST OF THE CENTER.

Example of WH forecast/advisory:
ZCZC MIATCMAT4
TTAA00 KNHC 190845
HURRICANE DANNY FORECAST/ADVISORY NUMBER 13
NATIONAL WEATHER SERVICE MIAMI FL AL0497
0900Z SAT JUL 19 1997

HURRICANE WARNINGS ARE IN EFFECT FROM GULFPORT MISSISSIPPI TO
APALACHICOLA FLORIDA. SMALL CRAFT SOUTHWEST OF GULFPORT SHOULD REMAIN
IN PORT UNTIL THE WINDS AND SEAS SUBSIDE.

HURRICANE CENTER LOCATED NEAR 30.2 N 88.0 W AT 19/0900Z POSITION ACCURATE
WITHIN 30 NM.

PRESENT MOVEMENT NEARLY STATIONARY

ESTIMATED MINIMUM CENTRAL PRESSURE 986 MB
MAX SUSTAINED WINDS 65 KTS WITH GUSTS TO 80 KT

64 KT	15NE	20SE	0SW	0NW
50 KT	20NE	30SE	30SW	0NW
34 KT	30 E	60SE	60SW	30NW
12FT SEAS	30NE	60SE	60SW	30NW

ALL QUADRANT RADII IN NAUTICAL MILES

FORECAST VALID 19/1800Z 30.2N 87.4W
MAX WIND   70 KT... GUSTS 85 KT

64 KT	20NE	20SE	20SW	20NW
50 KT	25NE	30SE	30SW	25NW
34 KT	30NE	75SE	75SW	30NW

CONVECTIVE OUTLOOK (AC)

A Convective Outlook (AC) is a national forecast of thunderstorms. There are two forecasts: Day 1 Convective Outlook (first 24 hours) and Day 2 Convective Outlook (next 24 hours). These forecasts describe areas in which there is a slight, moderate, or high risk of severe thunderstorms, as well as areas of general (non-severe) thunderstorms. The severe thunderstorm criteria are: Winds equal to or greater than 50 knots at the surface, or hail equal to or greater than ¾ inch in diameter at the surface, or tornadoes. The Refer to the Convective Outlook Chart (Section 12) for risk definitions. Forecast reasoning is also included in all ACs. Outlooks are produced by the Storm Prediction Center (SPC) located in Norman, OK. The times of issuance for Day 1 are 0600Z, 1300Z, 1630Z, 2000Z, and 0100Z. The initial Day 2 issuance is at 0830Z during standard time and 0730Z during daylight time. It is updated at 1730Z. The AC is a flight planning tool used to avoid thunderstorms.

Example:
MKC AC 291435
CONVECTIVE OUTLOOK… REF AFOS NMCGPH94O.
VALID 291500Z-301200Z

THERE IS A SLGT RISK OF SVR TSTMS TO THE RIGHT OF A LINE FROM 10 NE JAX 35 NNW
AYS AGS 15 E SPA 30 NE CLT 25 N FAY 30 ESE EWN.

GEN TSTMS ARE FCST TO THE RIGHT OF A LINE FROM 55 ESE YUM 30 NE IGM 15 S CDC 30
SW U24 25 ESE ELY 40 W P38 DRA 50 SW DRA 50 NW NID SAC 30 E ACV 25 E ONP 40 E BLI.

…SEVERE THUNDERSTORM FORECAST DISCUSSION…
.SERN U.S…

COOL FRONT CONTS SC/NC BORDER. VERY MOIST AND UNSTBL AMS ALONG AND S OF
FRONT E OF APLCHNS WITH CAPES TO REACH TO 4000 J/KG WITH AFTN HEATING.
ALTHOUGH WIND PROFILES ARE WK…COMB OF FRONTAL CNVGNC COUPLED WITH SEA
BREEZE FRONT WILL INITIATE PULSE SVR TSTMS VCNTY AND S OF FRONT THIS
AFTN/EVE. PRIMARY SVR EVENTS WILL BE WET DOWNBURST TO PUSH SWD FROM
CNTRL RCKYS EWD TO MID ATLC CST.   E OF APLCNS FRONT NOW LCTD VCNTY WND
DMG.

…GENERAL THUNDERSTORM FORECAST DISCUSSION…
…GULF CST AREA INTO SRN PLNS…

SFC FNT CURRENTLY LOCATED FM THE CAROLINAS WWD INTO PARTS OF OK WL CONT
TO SAG SLOWLY SWD ACRS THE SRN APLCNS/LWR MS VLY THRU THE REMAINDER OF
THE PD. S OF THE BNDRY…A VRY MOIST AMS RMNS IN PLACE AS DWPNTS ARE IN THE
MID TO UPR 70S. WHILE SOME CLDNS IS PRESENT ACRS THE AREA…SUF HEATING
SHOULD OCR TO ALLOW FOR MDT TO STG AMS DSTBLZN DURG THE LATE MRNG/ERY
AFTN. AS A RESULT…SFC BASED CAPE VALUES SHOULD BE AOA 2000 J/KG THIS AFTN.
BNDRYS FM OVERNIGHT CNVTN AS WELL AS SEA BREEZE CIRCULATIONS SHOULD BE
SUF TO INITIATE SCT TO NMRS TSTMS ACRS THE AREA. MID TO UPR LVL FLOW IS
RELATIVELY WK…SO THIS SUG ORGANIZED SVR TSTM ACTVTY IS NOT LIKELY.

SEVERE WEATHER WATCH BULLETINS (WWs) and ALERT MESSAGES (AWWs)

A Severe Weather Watch Bulletin (WW) defines areas of possible severe thunderstorms or tornado
activity. The bulletins are issued by the SPC in Norman, OK. WWs are unscheduled and are issued as
required.

A severe thunderstorm watch describes areas of expected severe thunderstorms. (Severe thunderstorm
criteria are ¾-inch hail or larger and/or wind gusts of 50 knots [58 mph] or greater.) A tornado watch
describes areas where the threat of tornadoes exists.

In order to alert the WFOs, CWSUs, FSSs, and other users, a preliminary notification of a watch called
the Alert Severe Weather Watch bulletin (AWW) is sent before the WW. (WFOs know this product as a
SAW).

Example of an AWW:
MKC AWW 011734
WW 75 TORNADO TX OK AR 011800Z-020000Z
AXIS..80 STATUTE MILES EAST AND WEST OF A LINE..60ESE DAL/DALLAS TX/ - 30 NW
ARG/ WALNUT RIDGE AR/
..AVIATION COORDS.. 70NM E/W /58W GGG - 25NW ARG/
HAIL SURFACE AND ALOFT..1 ¾ INCHES. WIND GUSTS..70 KNOTS. MAX TOPS TO 450.
MEAN WIND VECTOR 24045.

Soon after the AWW goes out, the actual watch bulletin itself is issued. A WW is in the following
format:

1. Type of severe weather watch, watch area, valid time period, type of severe weather possible, watch
    axis, meaning of a watch, and a statement that persons should be on the lookout for severe weather
2. Other watch information; i.e., references to previous watches
3. Phenomena, intensities, hail size, wind speed (knots), maximum CB tops, and estimated cell
    movement (mean wind vector)
4. Cause of severe weather
5. Information on updating ACs

Example of a WW:
BULLETIN - IMMEDIATE BROADCAST REQUESTED
TORNADO WATCH NUMBER 381
STORM PREDICTION CENTER NORMAN OK
556 PM CDT MON JUN 2 1997

THE STORM PREDICTON CENTER HAS ISSUED A TORNADO WATCH FOR PORTIONS OF

   NORTHEAST NEW MEXICO
   TEXAS PANHANDLE

EFFECTIVE THIS MONDAY NIGHT AND TUESDAY MORNING FROM 630 PM UNTIL
MIDNIGHT CDT.

TORNADOES… HAIL TO 2 ¾ INCHES IN DIAMETER… THUNDERSTORM WIND GUSTS TO 80
MPH… AND DANGEROUS LIGHTNING ARE POSSIBLE IN THESE AREAS.

THE TORNADO WATCH AREA IS ALONG AND 60 STATUTE MILES NORTH AND SOUTH OF
A LINE FROM 50 MILES SOUTHWEST OF RATON NEW MEXICO TO 50 MILES EAST OF
AMARILLO TEXAS.

REMEMBER… A TORNADO WATCH MEANS CONDITIONS ARE FAVORABLE FOR
TORNADOES AND SEVERE THUNDERSTORMS IN AND CLOSE TO THE WATCH AREA.
PERSONS IN THESE AREAS SHOULD BE ON THE LOOKOUT FOR THREATENING WEATHER
CONDITIONS AND LISTEN FOR LATER STATEMENTS AND POSSIBLE WARNINGS.

OTHER WATCH INFORMATION… CONTINUE… WW 378… WW 379… WW 380

DISCUSSION… THUNDERSTORMS ARE INCREASING OVER NE NM IN MOIST
SOUTHEASTERLY UPSLOPE FLOW. OUTFLOW BOUNDARY EXTENDS EASTWARD INTO
THE TEXAS PANHANDLE AND EXPECT STORMS TO MOVE ESE ALONG AND NORTH OF
THE BOUNDARY ON THE N EDGE OF THE CAP. VEERING WINDS WITH HEIGHT ALONG
WITH INCREASING MID LVL FLOW INDICATE A THREAT FOR SUPERCELLS.

AVIATION... TORNADOES AND A FEW SEVERE THUNDERSTORMS WITH HAIL SURFACE AND ALOFT TO 2 ¾ INCHES. EXTREME TURBULENCE AND SURFACE WIND GUSTS TO 70 KNOTS. A FEW CUMULONIMBI WITH MAXIMUM TOPS TO 550. MEANS STORM MOTION VECTOR 28025.

Status reports are issued as needed to show progress of storms and to delineate areas no longer under the threat of severe storm activity. Cancellation bulletins are issued when it becomes evident that no severe weather will develop or that storms have subsided and are no longer severe.

When tornadoes or severe thunderstorms have developed, the local WFO office will issue the warnings covering those areas.

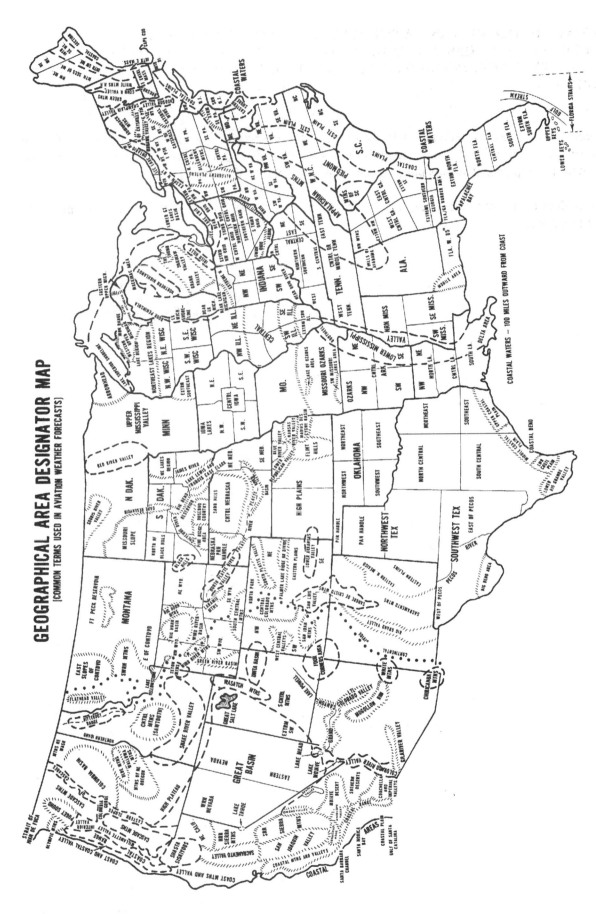

Figure 4-11. Geographical Areas and Terrain Features.

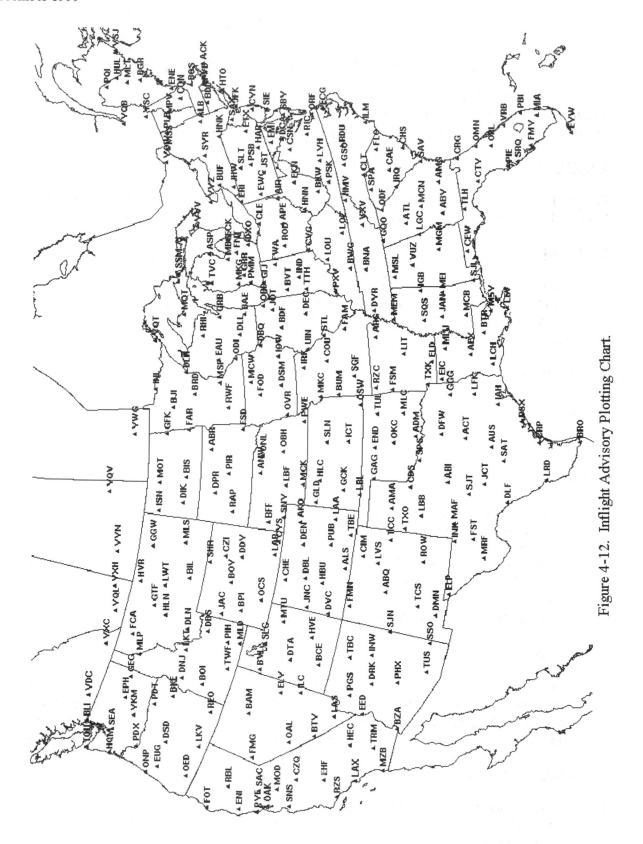

Figure 4-12. Inflight Advisory Plotting Chart.

Section 5
SURFACE ANALYSIS CHART

The surface analysis chart is a computer-generated chart, with frontal analysis by HPC forecasters, transmitted every 3 hours covering the contiguous 48 states and adjacent areas. Figure 5-1 is a surface analysis chart, and Figure 5-2 illustrates the symbols depicting fronts and pressure centers.

VALID TIME

Valid time of the chart corresponds to the time of the plotted observations. A date-time group in Universal Coordinated Time (UTC) tells the user when conditions portrayed on the chart occurred.

ISOBARS

Isobars are solid lines depicting the sea-level pressure pattern and are usually spaced at intervals of 4 millibar (mb), or hectoPascals (hPa) in metric units (1 millibar = 1 hectoPascal). Each isobar is labeled. For example, 1032 signifies 1,032.0 mb (hPa); 1000 signifies 1,000.0 mb (hPa); and 992 signifies 992.0 mb (hPa).

PRESSURE SYSTEMS

The letter "L" denotes a low pressure center, and the letter "H" denotes a high pressure center. The pressure of each center is indicated by a three- or four-digit number that is the central pressure in mb (hPa).

FRONTS

The analysis shows positions and types of fronts by the symbols in Figure 5-2. The symbols on the front indicate the type of front and point in the direction toward which the front is moving. If the front has arrowhead-shaped symbols, it is a cold front. If the front has half-moon symbols, it is a warm front. A three-digit number near a front classifies it as to type (see Table 5-1), intensity (see Table 5-2), and character (see Table 5-3). A bracket ([or]) before or after the number "points" to the front to which the number refers. For example, in Figure 5-1, the front extends from eastern Montana into central North Dakota south through South Dakota and Nebraska into northwestern Kansas. The front is labeled "027" which means a quasi-stationary front ("0" from Table 5-1); weak, little, or no change ("2" from Table 5-2); and with waves ("7" from Table 5-3).

Two short lines across a front indicate a change in classification. In figure 5-1, note that two lines cross the front in central Montana (adjacent to the Low). To the left of the Low the front is numbered "450" which is a cold front; moderate, little, or no change; and no specification. The front to the right of the Low is numbered "027" which is a quasi-stationary front; weak, little, or no change ; and with waves.

TROUGHS AND RIDGES

A trough of low pressure with significant weather will be depicted as a thick, dashed line running through the center of the trough and identified with the word "TROF." The symbol for a ridge of high pressure is very rarely, if at all, depicted (Figure 5-2).

OTHER INFORMATION

The observations from a number of stations are plotted on the chart to aid in analyzing and interpreting the surface weather features. These plotted observations are referred to as station models. There are two primary types of station models plotted on the chart. Those with a round station symbol are observations taken by observers. The locations with a square station symbol indicate the sky cover was determined by an automated system. Other plotting models that appear over water on the chart are data from ships, buoys, and offshore oil platforms. Figure 5-3 is an example of a station model that shows where the weather information is plotted. Figures 5-4 through Figure 5-7 help explain the decoding of the station model.

An outflow boundary will be depicted as a thick, dashed line with the word "OUTBNDY."

A dry line will be depicted as a line with unshaded pips or a through symbol. It will also be identified with the words "DRY LINE."

A legend is printed on each chart stating its name, valid date and valid time.

USING THE CHART

The surface analysis chart provides a ready means of locating pressure systems and fronts. It also gives an overview of winds, temperatures, and dew point temperatures at chart time. When using the chart, keep in mind that weather moves and conditions change. Using the surface analysis chart in conjunction with other information gives a more complete weather picture.

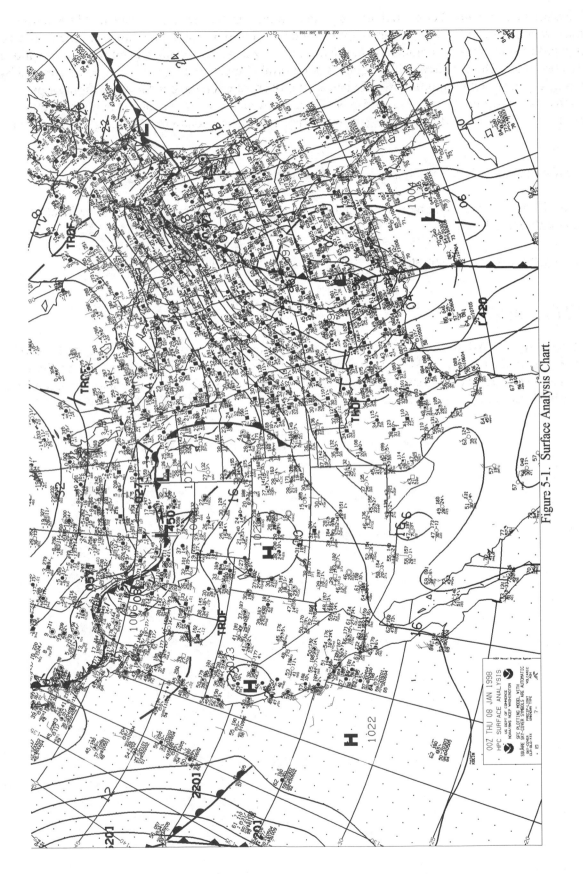

Figure 5-1. Surface Analysis Chart.

Color	Symbol	Description

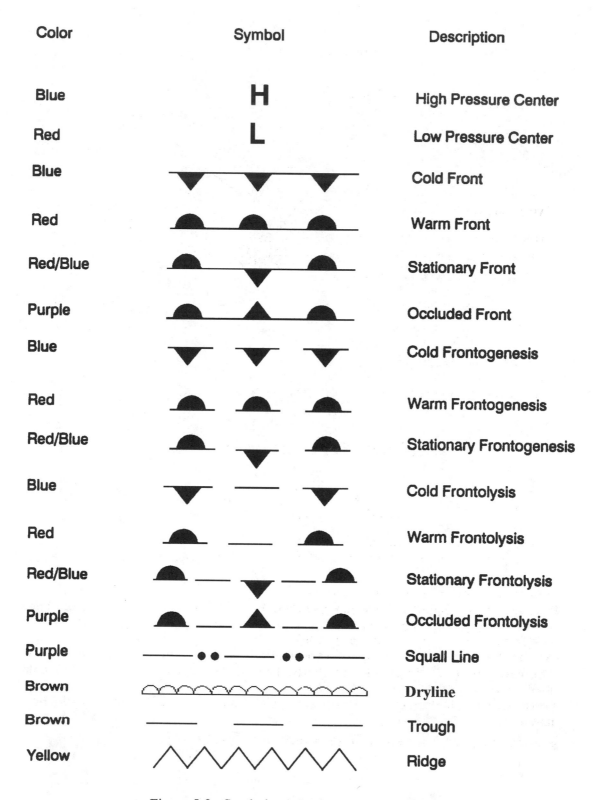

Blue	H	High Pressure Center
Red	L	Low Pressure Center
Blue		Cold Front
Red		Warm Front
Red/Blue		Stationary Front
Purple		Occluded Front
Blue		Cold Frontogenesis
Red		Warm Frontogenesis
Red/Blue		Stationary Frontogenesis
Blue		Cold Frontolysis
Red		Warm Frontolysis
Red/Blue		Stationary Frontolysis
Purple		Occluded Frontolysis
Purple		Squall Line
Brown		Dryline
Brown		Trough
Yellow		Ridge

Figure 5-2. Symbols on Surface Analysis Chart.

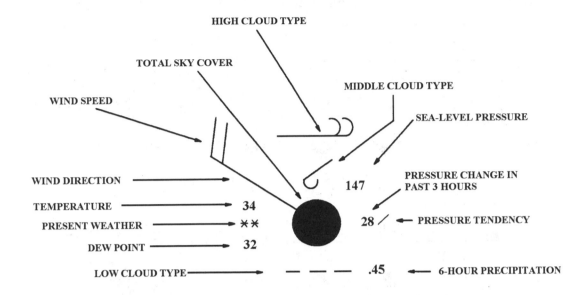

1. Total sky cover: Overcast.
2. Temperature: 34 degrees F, Dew Point: 32 degrees F.
3. Wind: From the northwest at 20 knots (relative to true north).

Examples of wind direction and speed

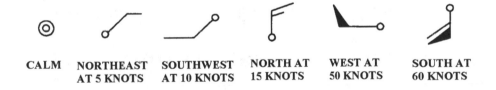

| CALM | NORTHEAST AT 5 KNOTS | SOUTHWEST AT 10 KNOTS | NORTH AT 15 KNOTS | WEST AT 50 KNOTS | SOUTH AT 60 KNOTS |

4. Present Weather: Continuous light snow.
5. Predominate low, middle, high cloud reported: Strato fractus or cumulus fractus of bad weather, altocumulus in patches, and dense cirrus.
6. Sea-level pressure: 1,014.7 millibars (mbs).
   NOTE: Pressure is always shown in three digits to nearest tenth of an mb. For 1,000 mbs or greater, prefix a "10" to the three digits. For less than 1,000 mbs, prefix a "9" to the three digits.
7. Pressure change in the past 3 hours: Increased steadily or unsteadily by 2.8 mbs. The actual change is in tenths of a mb.
8. 6 - hour precipitation in hundredths of an inch: 45 hundredths of an inch.

Figure 5-3. Station Model and Explanation.

Table 5-1. Type of Front

Code Figures	Descriptions
0	Quasi-stationary at surface
2	Warm front at surface
4	Cold front at surface
6	Occlusion
7	Instability line

Table 5-2. Intensity of Front

Code Figures	Descriptions
0	No specification
1	Weak, decreasing
2	Weak, little, or no change
3	Weak, increasing
4	Moderate, decreasing
5	Moderate, little, or no change
6	Moderate, increasing
7	Strong, decreasing
8	Strong, little, or no change
9	Strong, increasing

Table 5-3. Character of Front

Code Figures	Descriptions
0	No specification
5	Forming or existence expected
6	Quasi-stationary
7	With waves
8	Diffuse

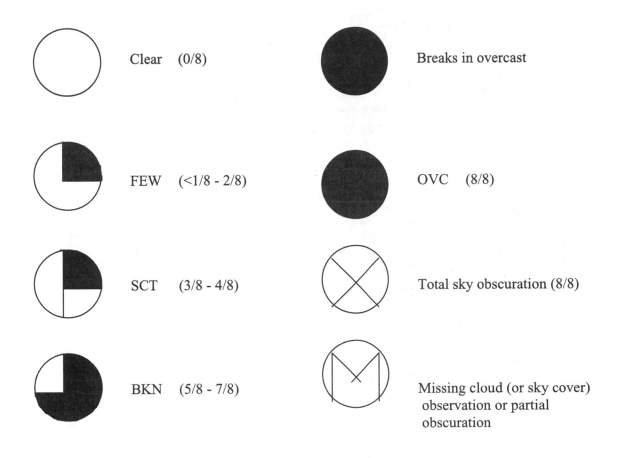

Figure 5-4. Sky Cover Symbols.

Description of Characteristic			
Primary Requirements	Additional Requirements	Graphic	Code Figure
Higher **Atmospheric pressure now higher than 3 hours ago.**	Increasing, then decreasing	 ⌒	0
	Increasing, then steady; or		
	Increasing, then increasing more slowly		1
	Increasing; steadily or unsteadily	/	2
	Decreasing; or steady, then increasing; or Increasing, then increasing more rapidly		3
Same **Atmospheric pressure now same as 3 hours ago.**	Increasing, then decreasing	 ⌒	0
	Steady	 —	4
	Decreasing, then increasing	 ⌄	5
Lower **Atmospheric pressure now lower than 3 hours ago.**	Decreasing, then increasing	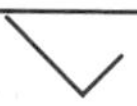 ⌄	5
	Decreasing, then steady; or		
	Decreasing, then decreasing more slowly		6
	Decreasing; steadily or unsteadily	\	7
	Steady; or increasing, then decreasing; or Decreasing, then decreasing more rapidly	 ⌃	8

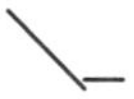

Figure 5-5. Pressure Tendencies.

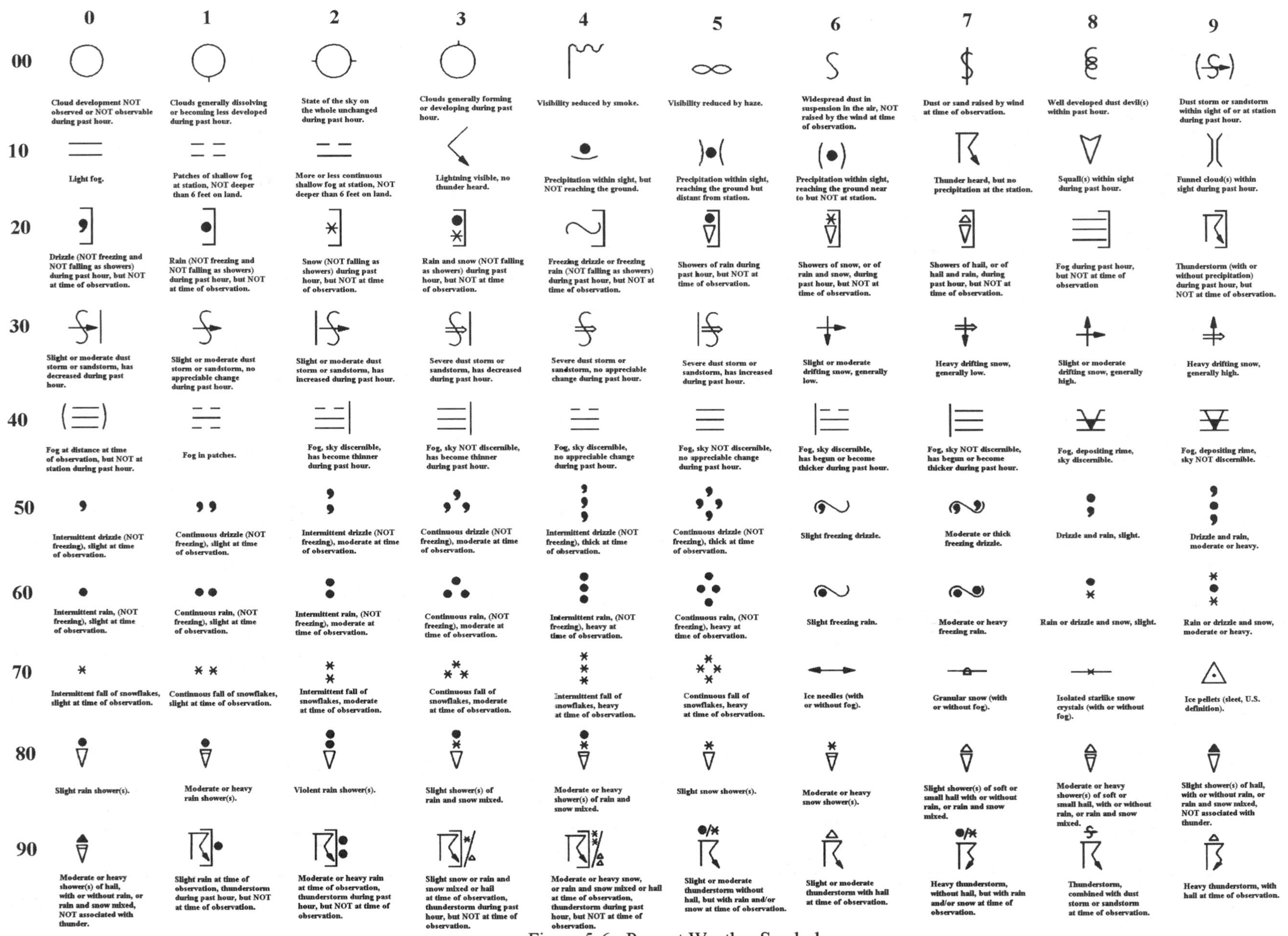

	0	1	2	3	4	5	6	7	8	9
00	Cloud development NOT observed or NOT observable during past hour.	Clouds generally dissolving or becoming less developed during past hour.	State of the sky on the whole unchanged during past hour.	Clouds generally forming or developing during past hour.	Visibility reduced by smoke.	Visibility reduced by haze.	Widespread dust in suspension in the air, NOT raised by the wind at time of observation.	Dust or sand raised by wind.	Well developed dust devil(s) within past hour.	Dust storm or sandstorm within sight of or at station during past hour.
10	Light fog.	Patches of shallow fog at station, NOT deeper than 6 feet on land.	More or less continuous shallow fog at station, NOT deeper than 6 feet on land.	Lightning visible, no thunder heard.	Precipitation within sight, NOT reaching the ground.	Precipitation within sight, reaching the ground but distant from station.	Precipitation within sight, reaching the ground near to but NOT at station.	Thunder heard, but no precipitation at the station.	Squall(s) within sight during past hour.	Funnel cloud(s) within sight during past hour.
20	Drizzle (NOT freezing and NOT falling as showers) during past hour, but NOT at time of observation.	Rain (NOT freezing and NOT falling as showers) during past hour, but NOT at time of observation.	Snow (NOT falling as showers) during past hour, but NOT at time of observation.	Rain and snow (NOT falling as showers) during past hour, but NOT at time of observation.	Freezing drizzle or freezing rain (NOT falling as showers) during past hour, but NOT at time of observation.	Showers of rain during past hour, but NOT at time of observation.	Showers of snow, or of rain and snow, during past hour, but NOT at time of observation.	Showers of hail, or of hail and rain, during past hour, but NOT at time of observation.	Fog during past hour, but NOT at time of observation.	Thunderstorm (with or without precipitation) during past hour, but NOT at time of observation.
30	Slight or moderate dust storm or sandstorm, has decreased during past hour.	Slight or moderate dust storm or sandstorm, no appreciable change during past hour.	Slight or moderate dust storm or sandstorm, has increased during past hour.	Severe dust storm or sandstorm, has decreased during past hour.	Severe dust storm or sandstorm, no appreciable change during past hour.	Severe dust storm or sandstorm, has increased during past hour.	Slight or moderate drifting snow, generally low.	Heavy drifting snow, generally low.	Slight or moderate drifting snow, generally high.	Heavy drifting snow, generally high.
40	Fog at distance at time of observation, but NOT at station during past hour.	Fog in patches.	Fog, sky discernible, has become thinner during past hour.	Fog, sky NOT discernible, has become thinner during past hour.	Fog, sky discernible, no appreciable change during past hour.	Fog, sky NOT discernible, no appreciable change during past hour.	Fog, sky discernible, has begun or become thicker during past hour.	Fog, sky NOT discernible, has begun or become thicker during past hour.	Fog, depositing rime, sky discernible.	Fog, depositing rime, sky NOT discernible.
50	Intermittent drizzle (NOT freezing), slight at time of observation.	Continuous drizzle (NOT freezing), slight at time of observation.	Intermittent drizzle (NOT freezing), moderate at time of observation.	Continuous drizzle (NOT freezing), moderate at time of observation.	Intermittent drizzle (NOT freezing), thick at time of observation.	Continuous drizzle (NOT freezing), thick at time of observation.	Slight freezing drizzle.	Moderate or thick freezing drizzle.	Drizzle and rain, slight.	Drizzle and rain, moderate or heavy.
60	Intermittent rain (NOT freezing), slight at time of observation.	Continuous rain (NOT freezing), slight at time of observation.	Intermittent rain (NOT freezing), moderate at time of observation.	Continuous rain (NOT freezing), moderate at time of observation.	Intermittent rain (NOT freezing), heavy at time of observation.	Continuous rain (NOT freezing), heavy at time of observation.	Slight freezing rain.	Moderate or heavy freezing rain.	Rain or drizzle and snow, slight.	Rain or drizzle and snow, moderate or heavy.
70	Intermittent fall of snowflakes, slight at time of observation.	Continuous fall of snowflakes, slight at time of observation.	Intermittent fall of snowflakes, moderate at time of observation.	Continuous fall of snowflakes, moderate at time of observation.	Intermittent fall of snowflakes, heavy at time of observation.	Continuous fall of snowflakes, heavy at time of observation.	Ice needles (with or without fog).	Granular snow (with or without fog).	Isolated starlike snow crystals (with or without fog).	Ice pellets (sleet, U.S. definition).
80	Slight rain shower(s).	Moderate or heavy rain shower(s).	Violent rain shower(s).	Slight shower(s) of rain and snow mixed.	Moderate or heavy shower(s) of rain and snow mixed.	Slight snow shower(s).	Moderate or heavy snow shower(s).	Slight shower(s) of soft or small hail with or without rain, or rain and snow mixed.	Moderate or heavy shower(s) of soft or small hail, with or without rain, or rain and snow mixed.	Slight shower(s) of hail, with or without rain, or rain and snow mixed, NOT associated with thunder.
90	Moderate or heavy shower(s) of hail, with or without rain, or rain and snow mixed, NOT associated with thunder.	Slight rain at time of observation, thunderstorm during past hour, but NOT at time of observation.	Moderate or heavy rain at time of observation, thunderstorm during past hour, but NOT at time of observation.	Slight snow or rain and snow mixed or hail at time of observation, thunderstorm during past hour, but NOT at time of observation.	Moderate or heavy snow, or rain and snow mixed or hail at time of observation, thunderstorm during past hour, but NOT at time of observation.	Slight or moderate thunderstorm without hail, but with rain and/or snow at time of observation.	Slight or moderate thunderstorm with hail at time of observation.	Heavy thunderstorm, without hail, but with rain and/or snow at time of observation.	Thunderstorm, combined with dust storm or sandstorm at time of observation.	Heavy thunderstorm with hail at time of observation.

Figure 5-6. Present Weather Symbols.

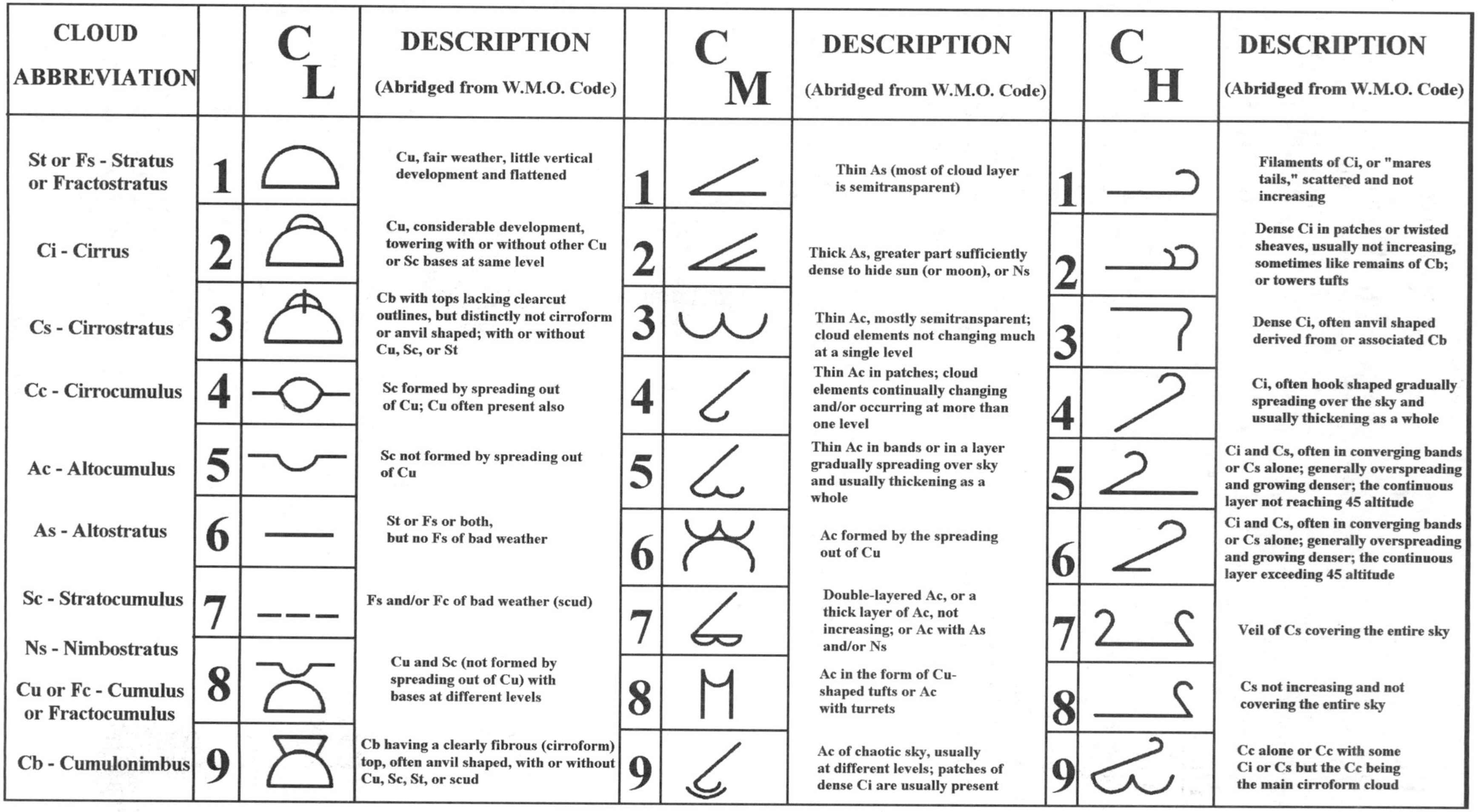

CLOUD ABBREVIATION

1. St or Fs - Stratus or Fractostratus
2. Ci - Cirrus
3. Cs - Cirrostratus
4. Cc - Cirrocumulus
5. Ac - Altocumulus
6. As - Altostratus
7. Sc - Stratocumulus
8. Ns - Nimbostratus
9. Cu or Fc - Cumulus or Fractocumulus
   Cb - Cumulonimbus

C_L — DESCRIPTION (Abridged from W.M.O. Code)

Code	Description
1	Cu, fair weather, little vertical development and flattened
2	Cu, considerable development, towering with or without other Cu or Sc bases at same level
3	Cb with tops lacking clearcut outlines, but distinctly not cirroform or anvil shaped; with or without Cu, Sc, or St
4	Sc formed by spreading out of Cu; Cu often present also
5	Sc not formed by spreading out of Cu
6	St or Fs or both, but no Fs of bad weather
7	Fs and/or Fc of bad weather (scud)
8	Cu and Sc (not formed by spreading out of Cu) with bases at different levels
9	Cb having a clearly fibrous (cirroform) top, often anvil shaped, with or without Cu, Sc, St, or scud

C_M — DESCRIPTION (Abridged from W.M.O. Code)

Code	Description
1	Thin As (most of cloud layer is semitransparent)
2	Thick As, greater part sufficiently dense to hide sun (or moon), or Ns
3	Thin Ac, mostly semitransparent; cloud elements not changing much at a single level
4	Thin Ac in patches; cloud elements continually changing and/or occurring at more than one level
5	Thin Ac in bands or in a layer gradually spreading over sky and usually thickening as a whole
6	Ac formed by the spreading out of Cu
7	Double-layered Ac, or a thick layer of Ac, not increasing; or Ac with As and/or Ns
8	Ac in the form of Cu-shaped tufts or Ac with turrets
9	Ac of chaotic sky, usually at different levels; patches of dense Ci are usually present

C_H — DESCRIPTION (Abridged from W.M.O. Code)

Code	Description
1	Filaments of Ci, or "mares tails," scattered and not increasing
2	Dense Ci in patches or twisted sheaves, usually not increasing, sometimes like remains of Cb; or towers tufts
3	Dense Ci, often anvil shaped derived from or associated Cb
4	Ci, often hook-shaped gradually spreading over the sky and usually thickening as a whole
5	Ci and Cs, often in converging bands or Cs alone; generally overspreading and growing denser; the continuous layer not reaching 45 altitude
6	Ci and Cs, often in converging bands or Cs alone; generally overspreading and growing denser; the continuous layer exceeding 45 altitude
7	Veil of Cs covering the entire sky
8	Cs not increasing and not covering the entire sky
9	Cc alone or Cc with some Ci or Cs but the Cc being the main cirroform cloud

Figure 5-7. Cloud Symbols.

Section 6
WEATHER DEPICTION CHART

The weather depiction chart, Figure 6-3, is computer-generated (with human frontal analysis) from METAR reports. The weather depiction chart gives a broad overview of the observed flying category conditions at the valid time of the chart. This chart begins at 01Z each day, is transmitted at 3-hours intervals, and is valid at the time of the plotted data.

PLOTTED DATA

Observations reported by both manual and automated observation locations provide the data for the chart. The right bracket ( ]) indicates the present weather information was obtained by an automated system only. The plotted data for each station are total sky cover, cloud height or ceiling, weather and obstructions to vision, and visibility. If the stations on the chart are crowded together, the weather, visibility, and cloud height may be moved up to 90 degrees around the station for better legibility. When reports are frequently updated, as at some automatic stations (every 20 minutes) or when the weather changes significantly, the observation used is the latest METAR received instead of using the one closest to the stated analysis time.

TOTAL SKY COVER

The amount of sky cover is shown by the station circle shaded as in Figure 6-1.

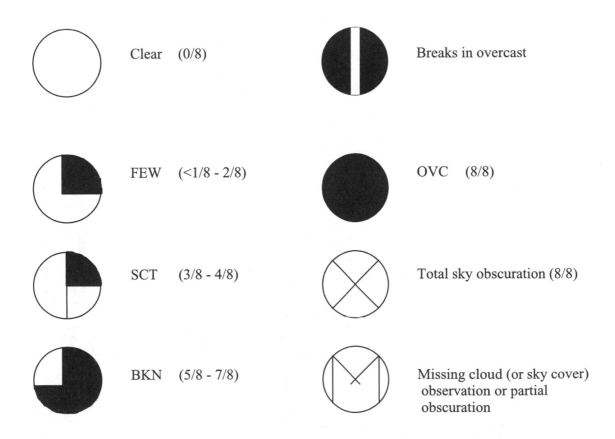

Figure 6-1. Total Sky Cover.

CLOUD HEIGHT

Cloud height above ground level (AGL) is entered under the station circle in hundreds of feet; the same as coded in a METAR report. If total sky cover at a station is scattered, the cloud height entered is the base of the lowest scattered cloud layer. If total sky cover is broken or greater at a station, the cloud height entered is the lowest broken or overcast cloud layer. A totally obscured sky is shown by the sky cover symbol "X" and is accompanied by the height entry of the obscuration (vertical visibility into the obscuration). A partially obscured sky without a cloud layer above, however, is not recognized by the computer program reading the METAR report. It cannot differentiate between a partial obscuration and a missing observation. Therefore, the computer program will enter an "M" in the sky cover circle for either occurrence. Consequently, the user will not know if the observation is missing or a partial obscuration is present. To obtain the most accurate information, the user must consult the METAR report for that specific station. A partially obscured sky with clouds above will have a cloud height entry for the cloud layer, but there will be no entry to indicate that there is a partial obscuration at the surface. So once again the user must consult the METAR report to obtain the most accurate information.

WEATHER AND OBSTRUCTIONS TO VISIBILITY

Weather and obstructions to visibility symbols are entered to the left of the station circle. Figure 5-6 explains most of the symbols used. When several types of weather and/or obstructions to visibility are reported at a station, the first one reported in the METAR would usually be the highest coded number in Figure 5-6. Also, for some stations that are not ordinarily plotted, the weather symbol is plotted only if the weather is significant, such as a thunderstorm.

VISIBILITY

When visibility is 5 miles or less, it is entered to the left of the weather or obstructions to vision symbol. Visibility is entered in statute miles and fractions of a mile.

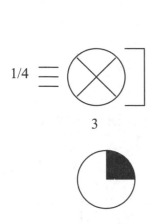

Total sky obscuration and the vertical visibility into the obscuration is 300 feet, visibility ¼, fog, bracket indicates fog was determined by an automated system

FEW sky coverage (no cloud height is indicated with FEW)

SCT sky coverage, cloud height 3,000' AGL, visibility 5 miles, haze

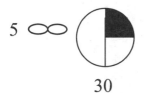

BKN sky coverage, ceiling height 2,000' AGL, visibility 3 miles, continuous rain

OVC sky coverage, ceiling height 500' AGL, visibility 1 mile, intermittent snow

SCT sky coverage, cloud height 25,000' AGL

BKN sky coverage, ceiling height 1,000' AGL, visibility 1½ mile, thunderstorm with rain shower

Missing cloud (or sky cover) observation or partial obscuration

Figure 6-2. Examples of Plotting on the Weather Depiction Chart.

ANALYSIS

The chart shows observed ceiling and visibility by categories as follows:

IFR - Ceiling less than 1,000 feet and/or visibility less than 3 miles; hatched area outlined by a smooth line.

MVFR (Marginal VFR) - Ceiling 1,000 to 3,000 feet inclusive and/or visibility 3 to 5 miles inclusive; non-hatched area outlined by a smooth line.

VFR - No ceiling or ceiling greater than 3,000 feet and visibility greater than 5 miles; not outlined.

The three categories are also explained in the lower right portion of the chart for quick reference. In addition, the chart shows fronts and troughs from the surface analysis for the preceding hour (with one exception being that fronts and troughs are omitted on the 10Z and 23Z charts). These features are depicted the same as the surface chart.

Because space on the chart is limited, only about half the METAR reports are plotted on the chart. The areas for each flight category are determined using all available reports whether or not they are plotted.

USING THE CHART

The weather depiction chart is an ideal place to begin preparing for a weather briefing and flight planning. From this chart, one can get a "bird's eye" view of areas of favorable and adverse weather conditions for chart time. This chart may not completely represent the en route conditions because of variations in terrain and possible weather occurring between reporting stations. Due to the delay between data and transmission time, changes in the weather could occur. One should update the chart with current METAR reports. After initially sizing up the general weather picture, final flight planning must consider forecasts, progs, and the latest pilot, radar, and surface weather reports.

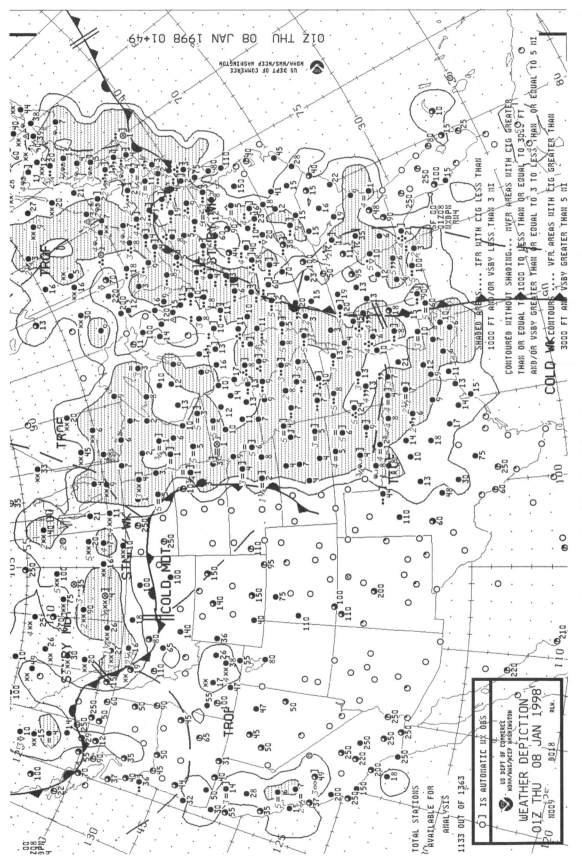

Figure 6-3. Weather Depiction Chart.

Section 7
RADAR SUMMARY CHART

A radar summary chart (Figure 7-1) is a computer-generated graphical display of a collection of automated radar weather reports (SDs). This chart displays areas of precipitation as well as information about type, intensity, configuration, coverage, echo top, and cell movement of precipitation. Severe weather watches are plotted if they are in effect when the chart is valid. The chart is available hourly with a valid time of H+35; i.e., 35 minutes past each hour. Figure 7-2 depicts the WSR-88D radar network from which the radar summary chart is developed.

ECHO (PRECIPITATION) TYPE

The types of precipitation are indicated on the chart by symbols located adjacent to precipitation areas on the chart. Table 7-1 lists the symbols used to denote types of precipitation. Note that these symbols do not reflect the change to METAR. Since the input data for the radar summary chart are the automated SDs, the type of precipitation is determined by computer models and is limited to the ones listed in Table 7-1.

Table 7-1 Key to Radar Chart

Symbols Used on Chart

Symbol	Meaning	Symbol	Meaning
R	Rain	↗ 35	Cell movement to the northeastat 35 knots
RW	Rain shower	LM	Little movement
S	Snow	WS999	Severe thunderstorm watch number 999
SW	Snow shower	WT210	Tornado watch number 210
T	Thunderstorm		
NA	Not available	SLD	8/10 or greater coverage in a line
NE	No echoes		
OM	Out for maintenancc	⟋	Line of echoes

INTENSITY

The intensity is obtained from the amount of energy returned to the radar from the target and is indicated on the chart by contours. Six precipitation intensity levels are reduced into three contour intervals as indicated in Table 7-2. In Figure 7-1, over central Montana is an area of precipitation depicted by one contour. The intensity of the precipitation area would be light to possibly moderate. Whether there is moderate precipitation in the area cannot be determined. However, what can be said is that the maximum intensity is definitely below heavy. When determining intensity levels from this chart, it is recommended that the maximum possible intensity be used. To determine the actual maximum intensity level, the SD for that time period should be examined. It should also be noted that intensity is coded for frozen precipitation (i.e., snow or snow showers). This is due to the fact that the WSR-88D is much more powerful and sensitive than previous radars. Finally, it is very important to remember that the intensity trend is no longer coded on the radar summary chart.

Table 7-2 Precipitation Intensities

Digit	Precipitation Intensity	Rainfall Rate in./hr. Stratiform	Rainfall Rate in./hr. Convective
1	Light	Less than 0.1	Less than 0.2
2	Moderate	0.1-0.5	0.2-1.1
3	Heavy	0.5-1.0	1.1-2.2
4	Very heavy	1.0-2.0	2.2-4.5
5	Intense	2.0-5.0	4.5-7.1
6	Extreme	More than 5.0	More than 7.1

450   Highest precipitation top in area in hundreds of feet MSL (45,000 feet MSL).

ECHO CONFIGURATION AND COVERAGE

The configuration is the arrangement of echoes. There are three designated arrangements: a LINE of echoes, an AREA of echoes, and an isolated CELL. (See Radar Weather Reports in Section 3 for definitions of the three configurations.)

Coverage is simply the area covered by echoes. All the hatched area inside the contours on the chart is considered to be covered by echoes. When the echoes are reported as a LINE, a line will be drawn through them on the chart. Where there is 8/10 coverage or more, the line is labeled as solid (SLD) at both ends. In the absence of this label, it can be assumed that there is less than 8/10 coverage. For example, in Figure 7-1, there is a solid line of thunderstorms with intense to extreme rain showers over central Georgia.

ECHO TOPS

Echo tops are obtained from both radar and, on occasion, satellite data and displayed for precipitation tops. Echo tops are the maximum heights of the precipitation in hundreds of feet MSL. They should be considered only as approximations because of radar wave propagation limitations. Tops are entered above a short line, with the top height displayed being the highest in the indicated area.

Examples:
220: maximum top 22,000 feet
500: Maximum top 50,000 feet

It is assumed that all precipitation displayed on the chart is reaching the surface. Some examples of top measurements in Figure 7-1 include a top of 15,000 feet MSL over northeast Washington; 23,000 feet over north-central Texas; and 32,000 feet MSL in central Georgia.

ECHO MOVEMENT

Individual cell movement is indicated by an arrow with the speed in knots entered as a number at the top of the arrow head. Little movement is identified by **LM**. For example, in Figure 7-1, the precipitation over north-central Texas is moving southwest at 8 knots. The precipitation in New England area is moving east-northeast at 25 knots. Line or area movement is no longer indicated on the chart.

SEVERE WEATHER WATCH AREAS

Severe weather watch areas are outlined by heavy dashed lines, usually in the form of a large rectangular box. There are two types - tornado watches and severe thunderstorm watches. Referring to Figure 7-1 and Table 7-1, the type of watch and the watch number are enclosed in a small rectangle and positioned as closely as possible to the northeast corner of the watch box. For example, in Figure 7-1, the boxed "WS0005" in northeast Georgia and western South Carolina is a severe thunderstorm watch and is the 5 [th] severe thunderstorm watch issued so far in the year. The watch number is also printed at the bottom of the chart (in Mexico) together with the issuance time and expiration time.

USING THE CHART

The radar summary chart aids in preflight planning by identifying general areas and movement of precipitation and/or thunderstorms. This chart displays drops or ice particles of precipitation size only; it does not display clouds and fog. Therefore, the absence of echoes does not guarantee clear weather, and cloud tops will most likely be higher than the tops of the precipitation echoes detected by radar. The chart must be used in conjunction with other charts, reports, and forecasts.

Examine chart notations carefully. Always determine location and movement of echoes. If echoes are anticipated near the planned route, take special note of echo intensity. Be sure to examine the chart for missing radar reports before assuming "no echoes present." For example, the Rapid City (RAP) radar report in western South Dakota is shown as "not available (NA)."

Suppose the planned flight route goes through an area of widely scattered thunderstorms in which no increase in area is anticipated. If these storms are separated by good VFR weather, they can be visually sighted and circumnavigated. However, widespread cloudiness may conceal the thunderstorms. To avoid these embedded thunderstorms, either use airborne radar or detour the area.
Remember that the radar summary chart is for preflight planning only and should be updated by current WSR-88D images and hourly reports. Once airborne, the pilot must evade individual storms by inflight

observations. This can be done by using visual sighting or airborne radar as well as by requesting radar echo information from Automated Flight Service Station (AFSS) Flight Watch. The AFSS Flight Watch has access to current WSR-88D imagery.

There can be an interpretation problem concerning an area of precipitation that is reported by more than one radar site. For example, station A may report RW with cell movement toward the northeast at 10 knots. For the same area, station B may be reporting TRW with cell movement toward the northeast at 30 knots. This difference in reports may be due to a different perspective and distance of the radar site from the area of echoes. The area may be moving away from station A and approaching station B. The rule of thumb is to use that plotted data associated with the area that presents the greatest hazard to aviation. In this case, the station B report would be used.

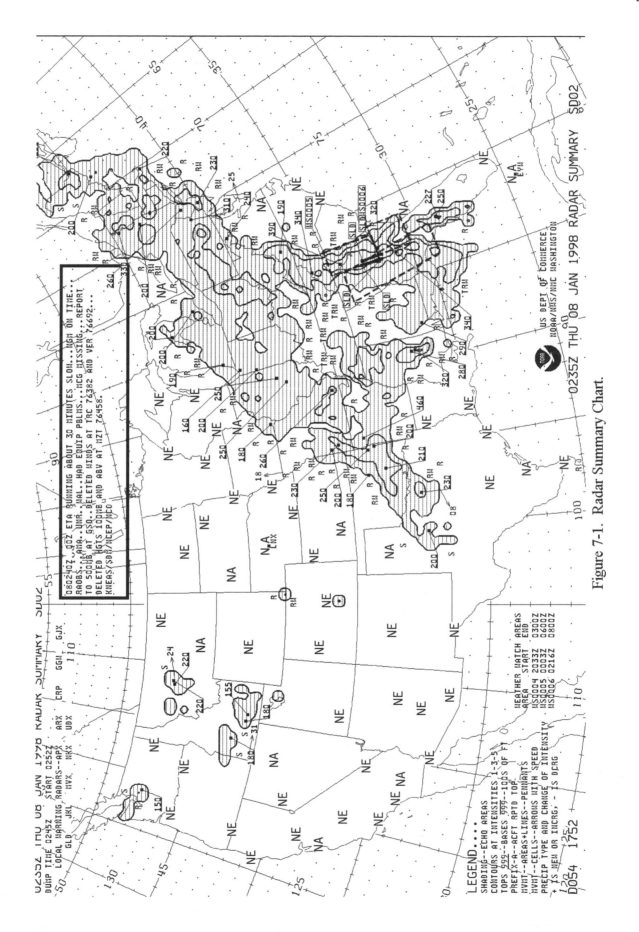

Figure 7-1. Radar Summary Chart.

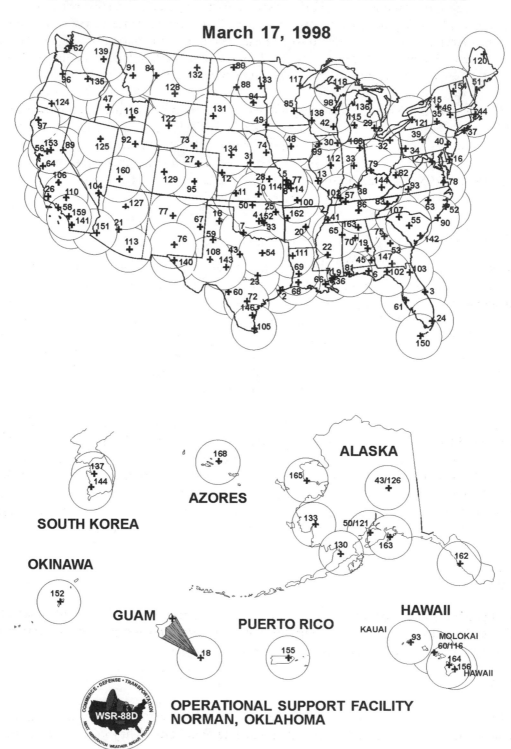

Figure 7-2. WSR-88D Radar Network.

Section 8
CONSTANT PRESSURE ANALYSIS CHARTS

Weather information for computer generated constant pressure charts is observed primarily by balloon-ascending radiosonde packages. Each package consists of weather instruments and a radio transmitter. During ascent instrument data are continuously transmitted to the observation station. Radiosondes are released at selected observational sites across the USA at 00Z and 12Z. The data collected from the radiosondes are used to prepare constant pressure charts twice a day.

Constant pressure charts are prepared for selected values of pressure and present weather information at various altitudes. The standard charts prepared are the 850 mb (hPa), 700 mb (hPa), 500 mb (hPa), 300 mb (hPa), 250 mb (hPa), and 200 mb (hPa) charts. Charts with higher pressures present information at lower altitudes, and charts with lower pressures present information at higher altitudes. Table 8-1 lists the general altitude (pressure altitude) of each constant pressure chart.

PLOTTED DATA

Data from each observation station are plotted around a station circle on each constant pressure chart. The circle identifies the station position. The data plotted on each chart are temperature, temperature-dew point spread, wind, height of the surface above sea level, and height change of the surface over the previous 12-hour period. The temperature and spread are in degrees Celsius, wind direction is relative to true north, wind speed is in knots, and height and height change are in meters. The station circle is shaded black when the spread is 5 degrees or less (moist atmosphere), and open when spread is more than 5 degrees (dry atmosphere). Figure 8-1 illustrates a station model of the radiosonde data plot. Table 8-2 gives station data plot examples for each constant pressure chart.

Aircraft and satellite observations are also used as information sources for constant pressure charts. A square is used to identify an aircraft reporting position. Data plotted are the flight level of the aircraft in hundreds of feet, temperature, wind, and time to the nearest hour UTC. A star is used to identify a satellite reporting position. Satellite information is determined by identifying cloud drift and height of cloud tops. Data plotted are the flight level in hundreds of feet, time to the nearest hour UTC, and wind. Aircraft and satellite data are plotted on the constant pressure chart closest to their reporting altitudes. Aircraft and satellite information are particularly useful over sparse radiosonde data areas.

ANALYSIS

All constant pressure charts contain analyses of height and temperature variations. Also, selected charts have analyses of wind speed variations. Variations of height are analyzed by contours, variations of temperature by isotherms, and variations of wind speed by isotachs.

CONTOURS

Contours are lines of constant height, in meters, which are referenced to mean sea level. Contours are used to map the height variations of surfaces that fluctuate in altitude. They identify and characterize pressure systems on constant pressure charts.

Contours are drawn as solid lines on constant pressure charts and are identified by a three-digit code located on each contour. To determine the contour height value, affix "zero" to the end of the code. For example, a contour with a "315" code on the 700 mb/hPa chart identifies the contour value as 3,150 meters. Also, affix a "one" in front of the code on all 200 mb/hPa contours and on 250 mb/hPa contours when the code begins with zero. For example, a contour with a "044" code on a 250 mb/hPa chart identifies the contour value as 10,440 meters.

The contour interval is the height difference between analyzed contours. A standard contour interval is used for each chart. The contour intervals are 30 meters for the 850 and 700 mb (hPa) charts, 60 meters for the 500 mb (hPa) chart, and 120 meters for the 300, 250, and 200 mb (hPa) charts.

The contour gradient is the distance between analyzed contours. Contour gradients identify slopes of surfaces that fluctuate in altitude. Strong gradients are closely spaced contours and identify steep slopes. Weak gradients are widely spaced contours and identify shallow slopes.

The contour analysis displays height patterns. Common types of patterns are lows, highs, troughs, and ridges. Contours have curvature for each of these patterns. Contour patterns can be further characterized by size and intensity. Size represents the breadth of a system. Sizes can range from large to small. A large pattern is generally more than 1,000 miles across, and a small pattern is less than 1,000 miles across. Intensities can range from strong to weak. Stronger systems are depicted by contours with stronger gradients and sharper curvatures. Weaker systems are depicted by contours with weaker gradients and weaker curvatures. For example, a chart may have a large, weak high, or a small, strong low.

Contour patterns on constant pressure charts can be interpreted the same as isobar patterns on the surface chart. For example, an area of low height is the same as an area of low pressure.

Winds respond to contour patterns and gradients. Wind directions parallel contours. In the Northern Hemisphere, when looking downwind, contours with relatively lower heights are to the left and contours with relatively higher heights are to the right. Thus, winds flow counterclockwise (cyclonically) around lows and clockwise (anticyclonically) around highs. (In the Southern Hemisphere these directions are reversed.) Winds that rotate are termed circulations. Wind speeds are faster with stronger gradients and slower with weaker gradients. In mountainous areas, winds are variable on pressure charts with altitudes at or below mountain crests. Contours have the effect of "channeling" the wind.

ISOTHERMS

Isotherms are lines of constant temperature. An isotherm separates colder air from warmer air. Isotherms are used to map temperature variations over a surface.

Isotherms are drawn as bold, dashed lines on constant pressure charts. Isotherm values are identified by a two-digit block on each line. The two digits are prefaced by "+" for above-freezing values as well as the zero isotherm and "-" for below-freezing values. Isotherms are drawn at 5-degree intervals on each chart. The zero isotherm separates above-freezing and below-freezing temperatures.

Isotherm gradients identify the magnitude of temperature variations. Strong gradients are closely spaced isotherms and identify large temperature variations. Weak gradients are loosely spaced isotherms and identify small temperature variations.

ISOTACHS

Isotachs are lines of constant wind speed. Isotachs separate higher wind speeds from lower wind speeds. Isotachs are used to map wind speed variations over a surface. Isotachs are analyzed on the 300, 250, and 200 mb (hPa) charts.

Isotachs are drawn as short, fine dashed lines. Isotach values are identified by a two- or three-digit number followed by a "K" located on each line. Isotachs are drawn at 20-knot intervals and begin at 10 knots.

Isotach gradients identify the magnitude of wind speed variations. Strong gradients are closely spaced isotachs and identify large wind speed variations. Weak gradients are loosely spaced isotachs and identify small wind speed variations.

Zones of very strong winds are highlighted by hatches. Hatched and unhatched areas are alternated at 40-knot intervals beginning with 70 knots. Areas between the 70- and 110-knot isotachs are hatched. Areas between the 110- and 150-knot isotachs are unhatched. This alternating pattern is continued until the strongest winds on the chart are highlighted. Highlighted isotachs assist in the identification of jet streams.

THREE-DIMENSIONAL ASPECTS

It is important to assess weather in both the horizontal and vertical dimensions. This not only applies to clouds, precipitation, and other significant conditions, but also pressure systems and winds. The characteristics of pressure systems vary horizontally and vertically in the atmosphere.

The horizontal distribution of pressure systems is depicted by the constant pressure charts and the surface chart (Section 5.) Pressure systems appear on each pressure chart as pressure patterns. Pressure charts identify and characterize pressure systems by their location, type, size, and intensity.

The vertical distribution of pressure systems must be determined by comparing pressure patterns on vertically adjacent pressure charts. For example, compare the surface chart with the 850 mb/hPa chart, 850 mb/hPa with 700 mb/hPa, and so forth. Changes of pressure patterns with height can be in the form of position, type, size, or intensity.

The three-dimensional assessment of pressure systems infers the assessment of the three-dimensional variations of wind.

USING THE CHARTS

Constant pressure charts are used to provide an overview of selected observed en route flying conditions. Use all pressure charts for a general overview of conditions.

Select the chart closest to the desired flight altitude for assessment of en route conditions. Review the winds along the route. Consider their direction and speed. For high altitude flights, identify jet stream positions. Note whether pressure patterns cause significant wind shifts or speed changes. Determine if these winds will be favorable or unfavorable (tailwind, headwind, crosswind.) Consider vertically adjacent charts and determine if a higher or lower altitude would have a more desirable en route wind. Interpolate winds between charts for flights between chart levels. Review other conditions along the route. Evaluate temperatures by identifying isotherm values and patterns. Evaluate areas with moist air and cloud potential by identifying station circles shaded black.

Consider the potential for hazardous flight conditions. Evaluate the potential for icing. Freezing temperatures and visible liquid forms of moisture produce icing. Evaluate the potential for turbulence. In addition to convective conditions and strong surface winds, turbulence is also associated with wind shear and mountain waves. Wind shear occurs with strong curved flow and speed shear. Strong lows and troughs and strong isotach gradients are indicators of strong shear. Vertical wind shear can be identified by comparing winds on vertically adjacent charts. Mountain waves are caused by strong perpendicular flow across mountain crests. Use winds on the pressure charts near mountain crest level to evaluate mountain wave potential.

Pressure patterns cause and characterize much of the weather. As a general rule, lows and troughs are associated with clouds and precipitation, while highs and ridges are associated with good weather. However, this rule is more complicated when pressure patterns change with height. Compare pressure pattern features on the various pressure charts with other weather charts, such as the weather depiction and radar summary charts. Note the association of pressure patterns on each chart with the weather.

Pressure systems, winds, temperature, and moisture change with time. For example, pressure systems move, change size, and change intensity. Forecast products predict these changes. Compare observed conditions with forecast conditions and be aware of these changes.

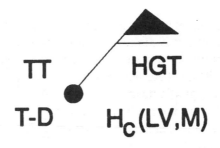

Code	Explanation
WIND:	Plotted wind direction and speed by symbol. Direction is to the nearest 10 degrees and speed is to the nearest 5 knots. (See Figure 5-3 for the explanation of the symbols.) If the direction or speed is missing, the wind symbol is omitted and an "M" is plotted in the H_c space. If speed is less than 3 knots, the wind is light and variable, the wind symbol is omitted, and an "LV" is plotted in the H_c space.
HGT:	Plotted height of the constant pressure surface in meters above mean sea level. (See Table 8-1 for decoding.) If data is missing, nothing is plotted in this position.
TT:	Plotted temperature to the nearest whole degree Celsius. A below-zero temperature is prefaced with a minus sign. Position is left blank if data is missing. A bracketed computer-generated temperature is plotted on the 850 mb/hPa chart in mountainous regions when stations have elevations above the 850 mb/hPa pressure level. If two temperatures are plotted, one above the other, the top temperature is used in the analysis.
T-D:	Plotted temperature-dew point spread to the nearest whole degree Celsius. An "X" is plotted when the air is extremely dry. The position is left blank when the information is missing.
H_c:	Plot of constant pressure surface height change which occurred during the previous 12 hours in tens of meters. For example, a +04 means the height of the surface rose 40 meters and a -12 means the height fell by 120 meters. H_c data is superseded by "LV" or "M" when pertinent.
CIRCLE:	Identifies station position. Shaded black when T-D spread is 5 degrees or less (moist). Unshaded when spread is more than 5 degrees.

Figure 8-1. Radiosonde Data Station Plot.

Table 8-1 Features of the Constant Pressure Charts - U.S.

Pressure (millibars/hectoPascals)	Pressure Altitude in feet (flight level)	Pressure Altitude in meters	Temperature/ Dew Point Spread	Isotachs	Contour Interval (meters)	Decode Station Height Plot		Examples of Station Height Plotting	
						Prefix to Plotted Value	Suffix to Plotted Value	Plotted	Height
850	5,000	1,500	yes	no	30	1	___	530	1,530
700	10,000	3,000	yes	no	30	2 or 3*	___	180	3,180
500	18,000	5,500	yes	no	60	_____	0	582	5,820
300	30,000	9,000	yes**	yes	120	_____	0	948	9,480
250	34,000	10,500	yes**	yes	120	1	0	063	10,630
200	39,000	12,000	yes**	yes	120	1	0	164	11,640

NOTE:
1. The pressure altitudes are rounded to the nearest 1,000 for feet and to the nearest 500 for meters.
2. All heights are above mean sea level (MSL).
3. * Prefix a "2" or "3," whichever brings the height closer to 3,000 meters.
4. ** Omitted when the air is too cold (temperature less than -41 degrees).

Table 8-2 Examples of Radiosonde Plotted Data

	22 479 ● 4 LV	09 ⟍ 129 17 -03	-19 ○ 558 X +03	-46 919 ○ +10	-55 037 ○ -01	-60 191 ○ M
	850 mb	**700 mb**	**500 mb**	**300 mb**	**250 mb**	**200 mb**
WIND	light and variable	010/20 KTS	210/60 KTS	270/25 KTS	240/30 KTS	missing
TT	22° C	9° C	-19° C	-46° C	-55° C	-60° C
T-D	4° C	17° C	>29° C	not plotted	not plotted	not plotted
DEW POINT	18° C	-8° C	dry	dry	dry	not plotted
HGT	1,479 m	3,129 m	5,580 m	9,190 m	10,370 m	11,910 m
H_c	not plotted	- 30 m	+ 30 m	+ 100 m	- 10 m	not plotted

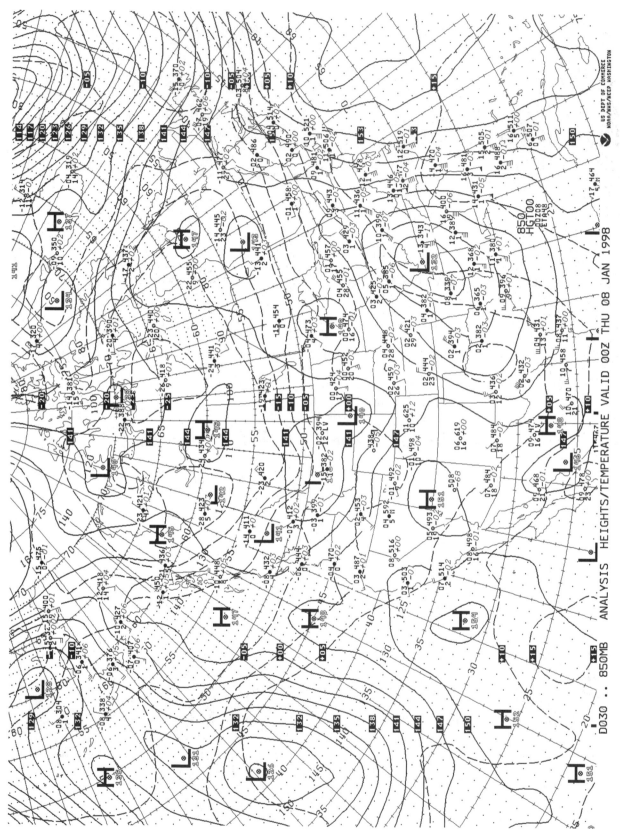

Figure 8-2. 850 Millibar/HectoPascal Analysis.

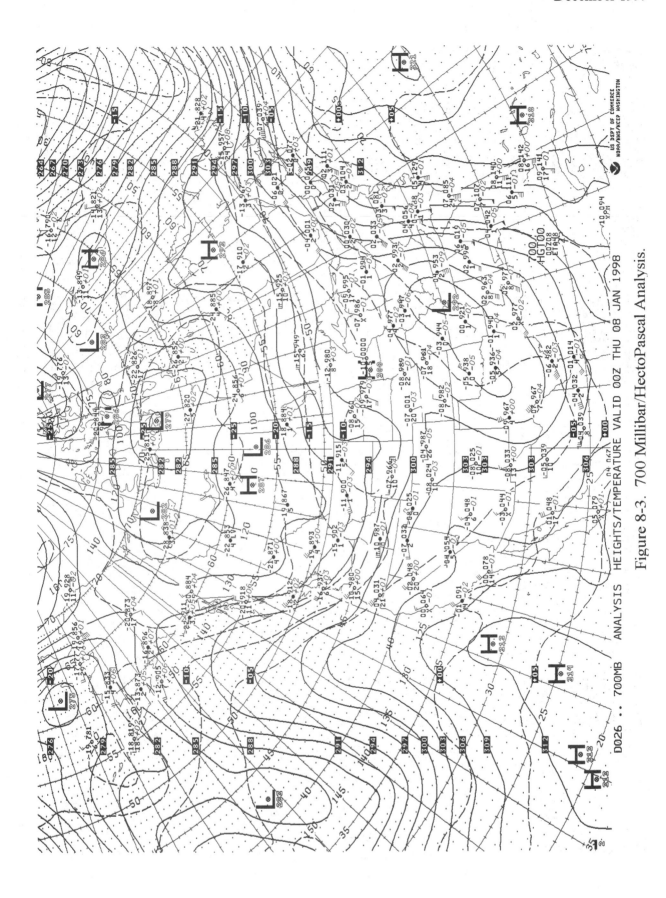

Figure 8-3. 700 Millibar/HectoPascal Analysis.

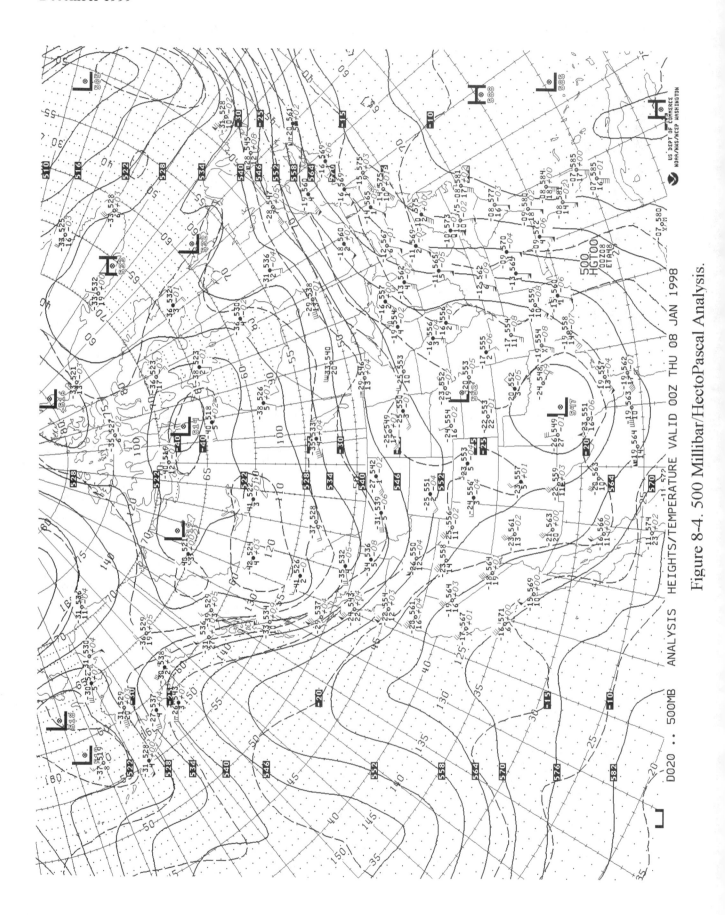

Figure 8-4. 500 Millibar/HectoPascal Analysis.

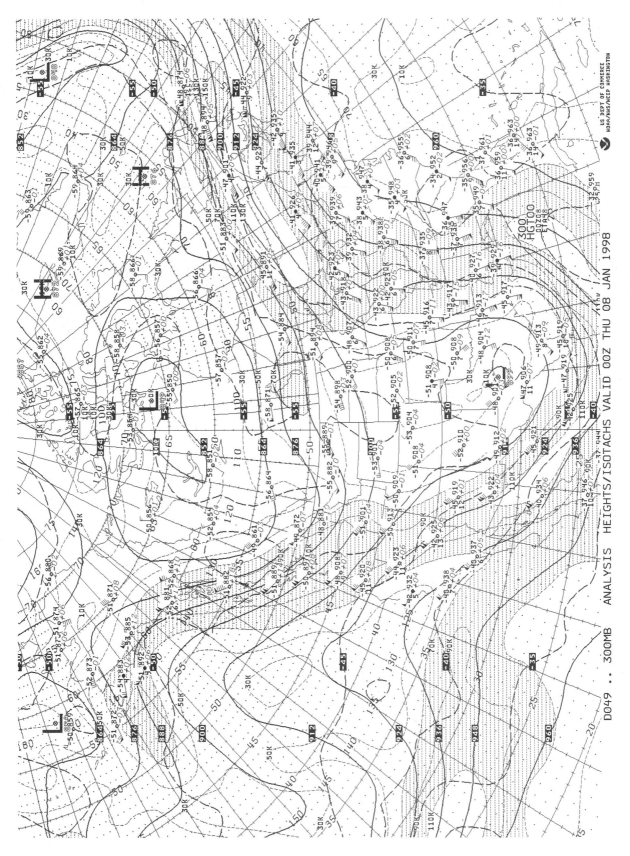

Figure 8-5. 300 Millibar/HectoPascal Analysis.

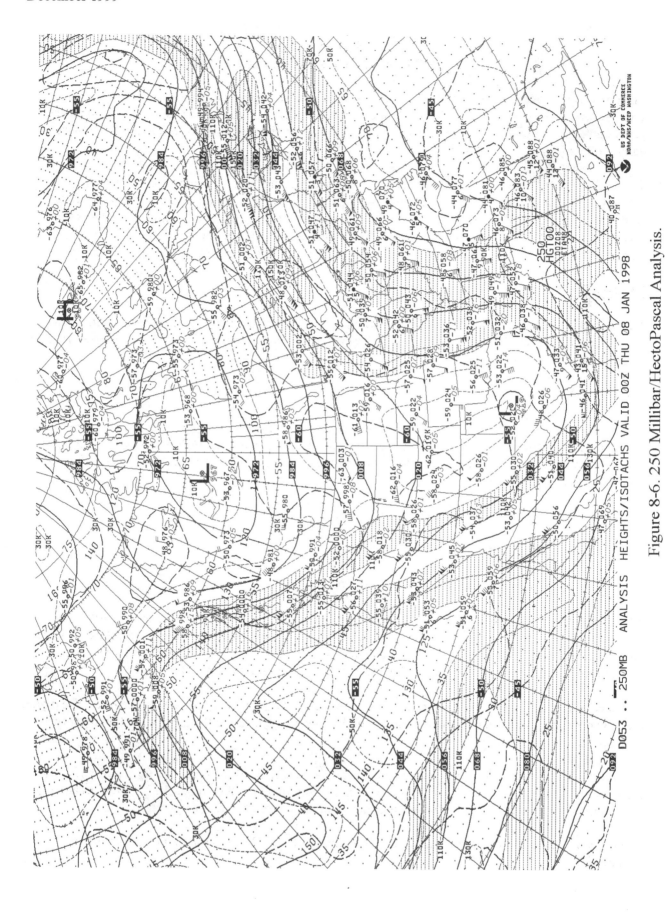

Figure 8-6. 250 Millibar/HectoPascal Analysis.

Section 9
COMPOSITE MOISTURE STABILITY CHART

The composite moisture stability chart (Figure 9-1) is a chart composed of four panels which depict stability, precipitable water, freezing level, and average relative humidity conditions. This computer-generated chart contains information obtained from upper-air observation data and is available twice daily with valid times of 00Z and 12Z.

The availability of upper-air data for analysis (on all panels) is indicated by the shape of the station symbols. Use the legend on the precipitable water panel (Figure 9-3) for the explanation of symbols common to all four panels. Mandatory levels referred to in the legend are the routinely used levels of surface; 1,000; 925; 850; 700; and 500 mb (hPa). Significant levels are nonroutine levels at which significant changes occur in the vertical profile of atmospheric properties during each observation.

STABILITY PANEL

The stability panel (Figure 9-2) is the upper left panel of the composite moisture stability chart. This panel contains two indexes that characterize the moisture and stability of the air. These indexes are the K index (KI) and the lifted index (LI).

K INDEX (KI)

The K index (KI) provides moisture and stability information. KI values range from high positive values to low negative values. A high positive KI implies moist and unstable air. A low or negative KI implies dry and stable air. KIs are considered high when values are at and above +20 and low when values are less than +20.

The KI is calculated by the summation of three terms:

$$KI = (850 \text{ mb/hPa temp} - 500 \text{ mb/hPa temp})$$
$$+ (850 \text{ Mb/hPa dew point})$$
$$- (700 \text{ mb/hPa temp/dew point spread})$$

The first term (850 mb/hPa temp - 500 mb/hPa temp) describes the vertical temperature profile. The term compares the temperature at about 5,000 feet mean sea level (MSL) with the temperature at about 18,000 feet MSL. The larger the temperature difference, the less stable the air, and the higher the KI. The smaller the temperature difference, the more stable the air, and the lower the KI.

The second term, 850 mb (hPa) dew point, is a measure of the quantity of low-level moisture. The higher the dew point, the higher the moisture, and the higher the KI. The lower the dew point, the lower the moisture, and the lower the KI.

The third term, 700 mb (hPa) temp/dew point spread, is a measure of the level of saturation at 700 mb (hPa). The smaller the spread, the higher the level of saturation, and the higher the KI. The greater the spread, the lower the level of saturation (drier air), and the lower the KI.

The KI can change significantly over a short time period of time due to temperature and moisture changes.

LIFTED INDEX (LI)

The lifted index (LI) is a common measure of atmospheric stability. The LI is obtained by hypothetically displacing a surface parcel upward to 500 mb (hPa) (about 18,000 feet MSL) and evaluating its stability at that level. A surface parcel is a small sample of air with representative surface temperature and moisture conditions. As the parcel is "lifted" it cools due to expansion. The temperature the parcel would have at 500 mb (hPa) is then subtracted from the actual (observed) 500 mb (hPa) temperature. This difference is the LI. LI values can be positive, negative, or zero. The LI does not identify the parcel's stability behavior at any of the intermediate altitudes between the surface and 500 mb (hPa).

A positive LI means a lifted surface parcel of air is stable. With a positive LI, the parcel would be colder and more dense than the surrounding air at 500 mb (hPa). A more dense parcel would resist upward motion. The stable surface parcel is like a rock at the bottom of a pool which, being more dense than the water, would resist being displaced upward. The more positive the LI, the more stable the air. Large positive values (+8 or greater) would indicate very stable air.

A negative LI means a lifted surface parcel of air is unstable. With a negative LI, the parcel would be warmer and less dense than the surrounding air at 500 mb (hPa). A parcel which is less dense than the surrounding air would continue to rise and possibly gain increasing upward speed until stabilizing at some higher altitude. The unstable surface parcel is like a cork at the bottom of a pool which, being less dense than the water, would accelerate upward to the surface of the pool. Large negative values (-6 or less) would indicate very unstable air.

A zero LI means a lifted surface parcel of air is neutrally stable. With a zero LI, the lifted parcel would have the same temperature and density as the air at 500 mb (hPa) and have a tendency to neither rise or sink. A neutrally stable parcel offers no resistance to vertical motion and, without further influence, would remain at the displaced level.

Temperature and moisture changes in the atmosphere change lifted index values. LIs decrease (become less stable) by increasing the surface temperatures, increasing surface dew points (moisture), and/or decreasing 500 mb (hPa) temperatures. Cold lows and troughs aloft with warm humid surface conditions tend to be associated with unstable air. LIs increase (become more stable) by decreasing surface temperatures, decreasing surface dew points, and/or increasing 500 mb (hPa) temperatures. Warm highs and ridges aloft with cool, dry surface conditions tend to be associated with stable air. Note that the LI can change considerably just by daytime heating and nighttime cooling of surface air. Daytime heating will decrease the LI, and nighttime cooling will increase the LI.

PLOTTED DATA

Figure 9-2 shows the two stability indexes that are computed for each upper-air station. The LI is plotted above the station symbol, and the KI is plotted below the symbol. Station circles are blackened for LI values of zero or less. An "M" indicates the value is missing

STABILITY ANALYSIS

The analysis is based on the LI only and highlights weakly stable and unstable areas. Solid lines are drawn for values of +4 and less at intervals of 4 (+4, 0, -4, -8, etc.).

USING THE PANEL

The KI and LI can be used in combination to assess moisture and stability properties of air masses. Air masses can be classified as moist and stable, moist and unstable, dry and stable, and dry and unstable. When used in combination, the KI, although containing stability information, is used primarily to classify moisture information, and the LI primarily to classify stability information. See Figure 9-2. Aberdeen, SD, has air characterized as dry and stable. The KI is 3 (dry) and the LI is 19 (stable). Melbourne, FL, is an example of dry and unstable air. The KI is 8 (dry) and the LI is -1 (unstable). Moist and unstable air is depicted at Key West, FL. The KI is 29 (moist) and the LI is -3 (unstable). The last example, Albany, NY, indicates moist and stable air. The KI is 31 (moist) and the LI is 15 (stable).

Moisture and stability properties of air masses characterize the weather. High KIs are associated with the potential for clouds and precipitation. Weather associated with high LIs and stable air are stratiform clouds and steady precipitation. Weather associated with low and negative LIs are unstable air, convective clouds, and showery precipitation.

The KI and LI can also be used to evaluate thunderstorm information. The KI is an indicator of the probability of thunderstorms (Table 9-1). Higher KIs imply higher probabilities. Lower KIs imply lower probabilities. The low and negative LIs are indicators of intensities of thunderstorms, if they occur. Higher negative LIs imply greater instability and stronger updrafts in thunderstorms. High positive LIs suggest little, if any, chance of thunderstorms.

Air masses classified with negative LIs do not always contain thunderstorms. This can occur for several reasons. Thunderstorm development is inhibited when a layer of stable air is located between the surface and 500 mb (hPa). This stable layer is referred to as a "cap." Inadequate amounts of moisture may also limit thunderstorm development in the presence of negative LIs. It is also possible to have a positive LI and still have thunderstorms develop. This can happen when a layer of air aloft located above stable surface air, such as above a front, is unstable and is sufficiently lifted, or if temperature and moisture conditions change rapidly and stabilities decrease.

Seasons affect the use of the KI regarding thunderstorm information. During the warmer seasons of spring, summer, and fall, a high KI generally indicates conditions are favorable for thunderstorms (Table 9-1). During winter with cold temperatures, fairly high values do not necessarily mean conditions are favorable for thunderstorms. Cold 850 mb (hPa) temperatures mean low dew points (low moisture.) The temperature profile term can generate high KI values, but low dew points may mean inadequate moisture to support thunderstorm development.

Table 9-1 Thunderstorm Potential

Lifted Index (LI)	Severe Potential	K Index (KI)	Thunderstorm Probability
0 to -2	Weak	< 15	near 0%
		15 - 19	20%
-2 to -6	Moderate	20 - 25	21% - 40%
		26 - 30	41% - 60%
≤ -6	Strong	31 - 35	61% - 80%
		36 - 40	81% - 90%
		> 40	near 100%

PRECIPITABLE WATER PANEL

The precipitable water panel (Figure 9-3) is the upper right panel of the composite moisture stability chart. This panel is an analysis of the quantity of water vapor in the atmosphere from the surface to the 500 mb (hPa) level (18,000 feet MSL). The quantity of water vapor is shown as precipitable water, which is the amount of liquid water that would result if all the water vapor were condensed.

Two constant factors affect precipitable water reports. Warm air is capable of holding higher quantities of water vapor than cold air. Therefore, warm air masses generally have more precipitable water than cold air masses. For example, precipitable water values are higher during summer months than during winter months. Also, high elevation stations have smaller vertical columns of air between surface and 500 mb (hPa) than low elevation stations. Therefore, higher elevation stations tend to have lower precipitable water than lower stations.

PLOTTED DATA

Precipitable water values are plotted above each station symbol to the nearest hundredth of an inch. The percent relative to normal for the month is plotted below the station symbol. Blackened circles indicate stations with precipitable water values of 1.00 inch or more. An "M" plotted above the station symbol indicates missing data.

ANALYSIS

Isopleths (lines of equal values) of precipitable water are drawn and labeled for every 0.25 inches with heavier isopleths drawn at 0.50-inch intervals.

USING THE CHART

This panel is used to determine the quantity of water vapor in the air between the surface and 500 mb (hPa) (18,000 feet MSL). Higher moisture content indicates "more fuel" for convective conditions. In Figure 9-3, Glasgow, MT, has a plot of ".22/100." This indicates that 22 hundreds of an inch of precipitable water is present, which is the average for the month. At Oklahoma City, OK, the ".72/196" indicates that 72 hundredths of an inch of precipitable water is present, which is 196 percent of normal (about double) for any day during this month. At Aberdeen, SD, the percent of normal value is not plotted due to insufficient climatological data.

FREEZING LEVEL PANEL

The freezing level panel (Figure 9-4) is the lower left panel of the composite moisture stability chart. This panel is an analysis of observed freezing levels. The freezing level is the height above MSL at which the temperature is zero degrees Celsius.

Freezing levels are affected by air mass temperatures. Colder air masses have lower freezing levels, and warmer air masses have higher freezing levels. Freezing levels change with the movement of contrasting cold and warm air masses. For example, freezing levels tend to lower behind cold fronts and rise ahead of warm fronts.

Generally, a station has one freezing level. Relative to the freezing level, the lower levels have above-freezing temperatures, and the upper levels have below-freezing temperatures. During very cold periods, all temperatures over a station may be below freezing and there would be no freezing level.

During colder periods of the year, and with certain weather systems such as fronts, stations may have more than one freezing level. There would be several layers of air with alternating above-freezing and below-freezing temperatures. A report from such a station would contain multiple freezing levels. Table 9-2 illustrates a vertical temperature profile drawn relative to zero degrees Celsius which contains multiple freezing levels. In this table there are two layers with above-freezing temperatures and two layers with below-freezing temperatures. Above-freezing layers extend from the surface to 3,000 feet MSL and from 6,000 to 9,000 feet MSL. Below-freezing layers extend from 3,000 to 6,000 feet MSL and above 9,000 feet MSL.

Table 9-2 Vertical Temperature Profile with Freezing Levels

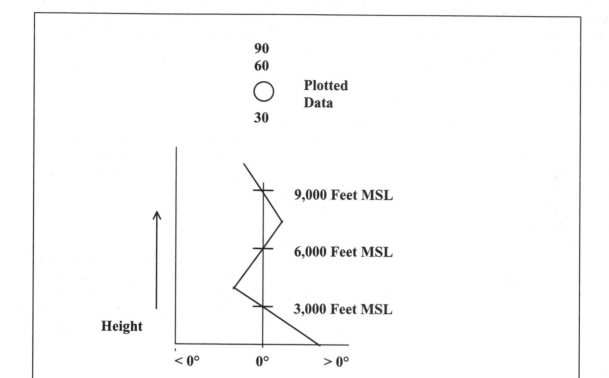

PLOTTED DATA

Observed freezing levels are plotted on the chart in hundreds of feet MSL. Multiple freezing level events have plots for each freezing level. BF is plotted on the chart to indicate below-freezing temperatures at the surface. "M" indicates missing data. Note in Table 9-2 the freezing level plots associated with the illustrated vertical temperature profile. Table 9-3 provides examples of several station plots for various types of freezing level conditions.

ANALYSIS

Freezing levels are analyzed with contours (lines of constant height) and are drawn as solid lines. The lines are drawn with intervals of 4,000 feet beginning with 4,000 feet. Multiple freezing levels are analyzed for the lowest freezing level. Contours are labeled in hundreds of feet MSL. The surface freezing level is drawn and labeled as the 32-degree Fahrenheit (0° C)isotherm. The surface freezing level line encloses stations with BF data plots.

USING THE PANEL

The freezing level chart is used to assess freezing level heights and their values relative to flight profiles. In Figure 9-4, Salt Lake City, UT, is an example where all temperatures above the station were below freezing (below 0° C or 32° F.) Lake Charles, LA, depicts a single freezing level at 11,500 feet MSL. North Platte, NE, is an example of multiple freezing levels. The temperatures were below freezing at the surface but warmed to above freezing between 4,400 and 6,100 feet MSL. Above 6,100 feet MSL the temperatures were again below freezing.

In areas with single freezing levels, flights above the freezing level will be in below-freezing temperatures, and flights below the freezing level will be in above-freezing temperatures. In areas with multiple freezing levels, there are multiple layers of above- and below-freezing temperatures. According to Figure 9-4, a flight en route from Seattle, WA, to Portland, OR, at 7,000 feet would be flying above the freezing level and in below-freezing temperatures. A flight en route at 7,000 feet from Atlanta, GA, to Washington, DC, would be flying below the freezing level and in above-freezing temperatures.

Special care must be exercised to properly identify the altitudes of layers with above and below freezing temperatures when there is a potential for icing conditions.

Table 9-3 Plotting Freezing Levels

Plotted	Interpretation
◯ BF	Entire observation is below freezing (0 degrees Celsius).
28 ✳	Freezing level is at 2,800 feet MSL; temperatures below freezing above 2,800 feet MSL. All significant levels are missing.
120 ☐	Freezing level at 12,000 feet; temperatures above 12,000 feet are below freezing. Some mandatory levels are missing.
110 51 ◯ BF	Temperatures are below freezing from the surface to 5,100 feet; above freezing from 5,100 to 11,000 feet MSL; and below freezing above 11,000 feet MSL.
90 34 ◯ 3	Lowest freezing level is at 300 feet; below freezing from 300 feet to 3,400 feet; above freezing from 3,400 to 9,000 feet; and below freezing above 9,000 feet.
M ◯	Data is missing.

AVERAGE RELATIVE HUMIDITY PANEL

The average relative humidity panel (Figure 9-5) is the lower right panel of the composite moisture stability chart. This panel is an analysis of the average relative humidity for the layer surface to 500 mb (hPa).

Relative humidity is the ratio of the quantity of water vapor in a sample of air compared to the air's capacity to hold water vapor expressed in percent. The air's capacity to hold water vapor depends primarily on its temperature and, to a lesser extent, its pressure. Warm air can hold more water vapor than cold air. Air at lower pressure can hold more water vapor than air at higher pressure.

Average relative humidities of the layer are changed primarily by vertical motion of air. Upward motion increases relative humidities, and downward motion decreases relative humidities.

Relative humidity is an indicator of the degree to which air is saturated. Air is saturated when it contains all of the water vapor it can hold. High relative humidities (moist air) identify air which is at or close to saturation. Air with high relative humidites often contain clouds and may produce precipitation. Low relative humidities (dry air) identify air that is not close to saturation. Low relative humidity air tends to be free of clouds.

PLOTTED DATA

The average relative humidity is plotted above each station symbol. Blackened circles indicate stations with humidities of 50 percent and higher. An "M" indicates the value is missing.

ANALYSIS

Isopleths of relative humidity, called isohumes, are drawn and labeled every 10 percent, with more heavily shaded isohumes drawn for values of 10, 50, and 90 percent.

USING THE PANEL

This panel is used to determine the average relative humidity of air from the surface to 500 mb (hPa). Areas with high average relative humidities have a higher probability of thick clouds and possibly precipitation. Areas with low average relative humidities have a lower probability of thick clouds, although shallow cloud layers may be present. Weather-producing systems, such as lows and fronts, which are moving into areas with high average relative humidities have a high probability of developing clouds and precipitation. Significant values of average relative humidities which support the possibility of developing clouds and precipitation are 50% and higher with unstable air, and 70% and higher for stable air. Weather-producing systems affecting areas with low average relative humidities, 30% and less, may produce only a few clouds, if any. According to Figure 9-5, much of Arkansas has very moist air with average relative humidities greater than 90%, while western Arizona has dry air with average relative humidities less than 30%.

High values of relative humidity do not necessarily mean high values of water vapor content (precipitable water). For example in Figure 9-3, Oakland, CA, had less water vapor content than Miami, FL (.64 and 1.43 respectively). However, in Figure 9-5, the average relative humidities are the same for both stations. If rain were falling at both stations, the result would likely be lighter precipitation totals for Oakland.

USING THE CHART

This chart is used to identify the distribution of moisture, stability, and freezing level properties of the atmosphere. These properties and their association with weather systems provide important insights into existing and forecast weather conditions as well as possible aviation weather hazards.

Generally these properties tend to move with the associated weather systems, such as lows, highs, and fronts. Contrasting property values within weather systems are redistributed relative to the systems by advecting winds. Also, changes in property values relative to the systems can occur as a result of development and dissipation processes. In some instances property values will remain stationary relative to geographical features, such as mountains and coastal regions.

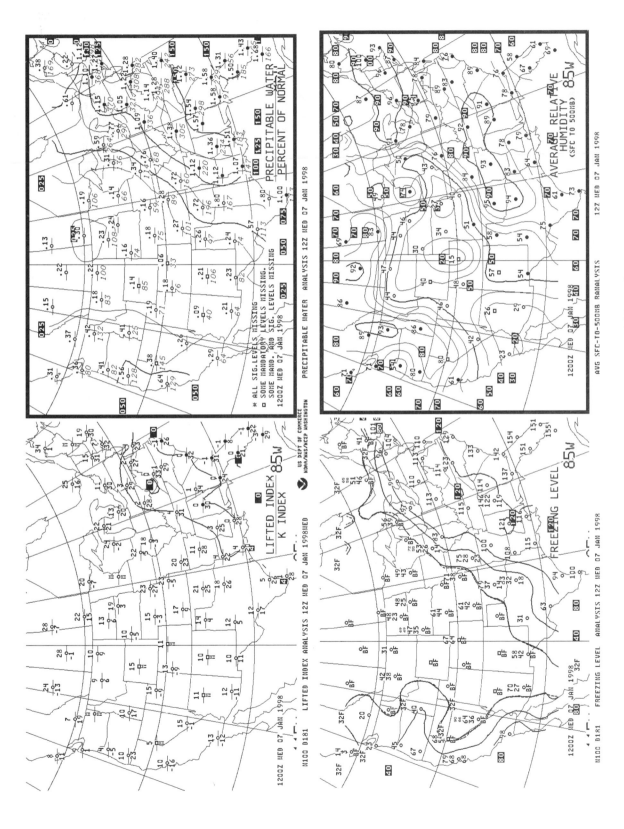

Figure 9-1. Composite Moisture Stability Chart.

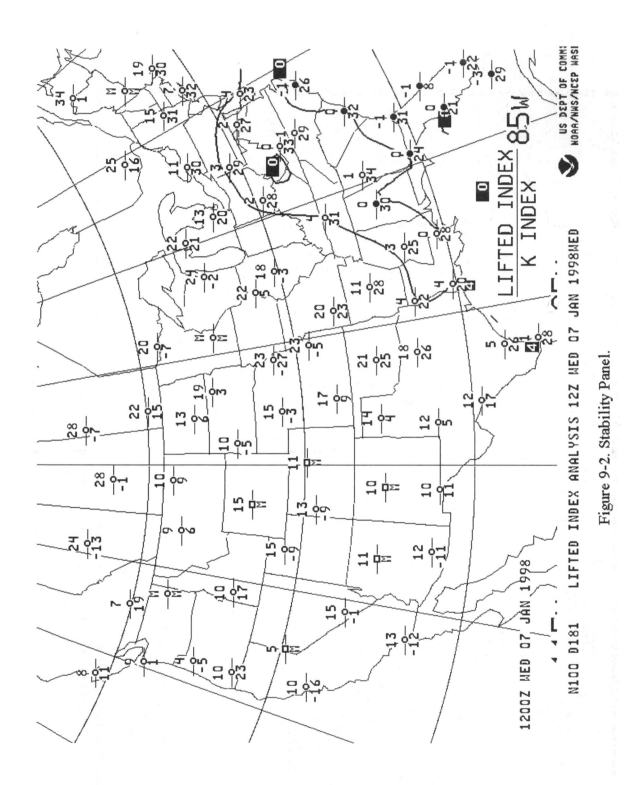

Figure 9-2. Stability Panel.

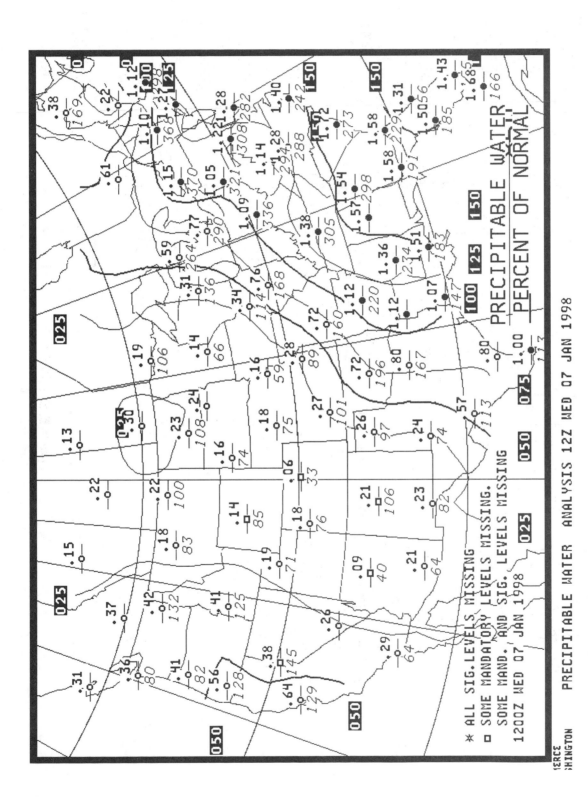

Figure 9-3. Precipitable Water Panel.

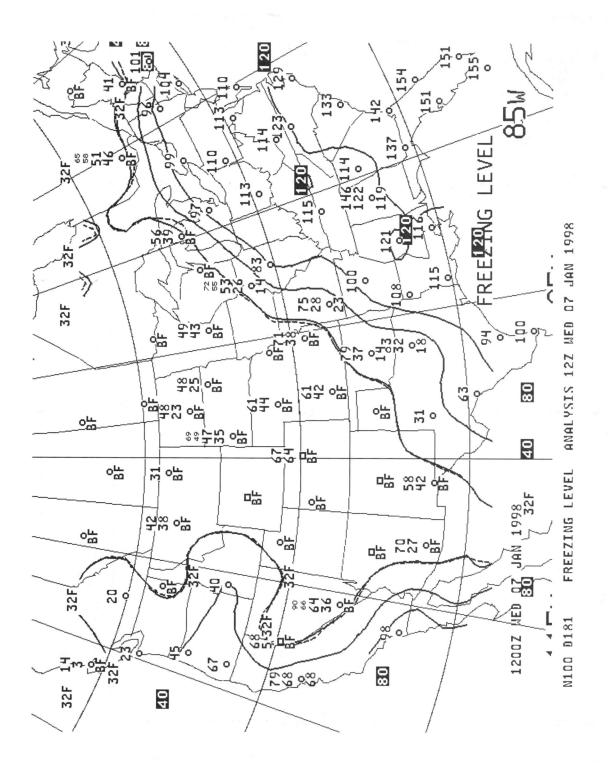

Figure 9-4. Freezing Level Panel.

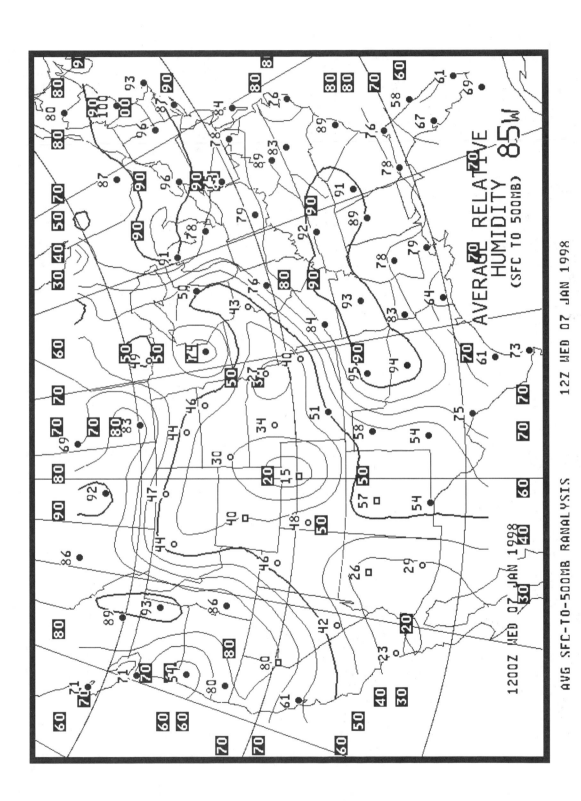

Figure 9-5. Average Relative Humidity Panel.

Section 10
WINDS AND TEMPERATURES ALOFT CHARTS

Winds aloft charts, both forecast and observed, are computer-generated products. The forecast winds aloft charts also contain forecast temperatures aloft.

FORECAST WINDS AND TEMPERATURES ALOFT (FD)

Forecast winds and temperatures aloft (FD) charts are prepared for eight levels on eight separate panels. The levels are 6,000; 9,000; 12,000; 18,000; 24,000; 30,000; 34,000; and 39,000 feet MSL. They are available daily, and the 12-hour progs are valid at 1200Z and 0000Z. A legend on each panel shows the valid time and the level of the panel. Levels below 18,000 feet are in true altitude, and levels 18,000 feet and above are in pressure altitude. Figure 10-1 shows examples from a winds and temperatures aloft forecast chart. Figure 10-2 provides a closer view of the winds and temperature aloft forecast chart. Temperature is in whole degrees Celsius for each forecast point and is entered above and to the right of the station circle. Arrows with pennants and barbs, similar to those used on the surface map, show wind direction and speed. Wind direction is drawn to the nearest 10 degrees with the second digit of the coded direction entered at the outer end of the arrow. To determine wind direction, obtain the general direction from the arrow, and then use the digit to determine the direction to the nearest 10 degrees. For example, a wind in the northwest quadrant with a digit of "3" indicates 330 degrees. A calm or light and variable wind is shown by "99" entered to the lower left of the station circle. See Table 10-1 for examples of plotted temperatures and winds with their interpretations.

Table 10-1 Plotted Winds and Temperatures

Plotted	Interpretation

12 ⟍ **6**

12 degrees Celsius, wind 060 degrees at 5 knots

3

3 degrees Celsius, wind 160 degrees at 25 knots

6

0

5

0 degrees Celsius, wind 250 degrees at 15 knots

-09

6

-9 degrees Celsius, wind 260 degrees at 50 knots

6

-47

-47 degrees Celsius, wind 360 degrees at 115 knots

-11

99

-11 degrees Celsius, wind calm or light and variable

OBSERVED WINDS ALOFT

Charts of observed winds for selected levels are sent twice daily on a four-panel chart valid at 1200Z and 0000Z. The chart depicts winds and temperatures at the second standard level, 14,000, 24,000, and 34,000 feet. Figure 10-3 is an example of this chart, and Figure 10-4 is an example of one of the panels. Wind direction and speed are shown by arrows, the same as on the forecast charts. A calm or light and variable wind is shown as "LV" and a missing wind as "M," both plotted to the lower right of the station circle. The station circle is filled in when the reported temperature/dew point spread is 5 degrees Celsius or less. Observed temperatures are included on the upper two panels of this chart (24,000 feet and 34,000 feet). A dotted bracket around the temperature means a calculated temperature.

The second standard level (Figure 10-3) for a reporting station is found between 1,000 and 2,000 feet above the surface, depending on the station elevation. The second standard level is used to determine low-level wind shear and frictional effects on lower atmosphere winds. To compute the second standard level, find the next thousand-foot level above the station elevation and add 1,000 feet to that level. For example, the next thousand-foot level above Oklahoma City, OK, (station elevation 1,290 feet MSL) is 2,000 feet MSL. The second standard level for Oklahoma City, OK, (2,000 feet + 1,000 feet) is 3,000 feet MSL or 1,710 feet AGL.

For example:

Station:			
Amarillo, TX	Bismarck, ND	Topeka, KS	Key West, FL

Station elevation:

3,604 MSL	1,677 MSL	879 MSL	0 MSL

Next thousand-foot level above station:

4,000 MSL	2,000 MSL	1,000 MSL	1,000 MSL
+1,000	+1,000	+1,000	+1,000
————	————	————	————

Second standard level:

5,000 MSL	3,000 MSL	2,000 MSL	2,000 MSL
or	or	or	or
1,396 AGL	1,323 AGL	1,121 AGL	2,000 AGL

Note that the 12,000 foot MSL panel is true altitude, while the 24,000 and 34,000 feet MSL panels are in pressure altitude. (See Figure 10-1.)

USING THE CHARTS

The use of the winds aloft chart is to determine winds at a proposed flight altitude or to select the best altitude for a proposed flight. Temperatures also can be determined from the forecast charts. Interpolation must be used to determine winds and temperatures at a level between charts and data when the time period is other than the valid time of the chart.

Forecast winds are generally preferable to observed winds since they are more relevant to flight time. Although observed winds are 5 to 8 hours old when received, they still can be a useful reference to check for gross errors on the 12-hour prog.

INTERNATIONAL FLIGHTS

Computer-generated forecast charts of winds and temperatures aloft are available for international flights at specified levels. The U.S. National Centers of Environmental Prediction (NCEP), near Washington D.C., is a component of the World Area Forecast System (WAFS). NCEP is designated in the WAFS as both a World Area Forecast Center and a Regional Area Forecast Center (RAFC). Its main function as a World Area Forecast Center is to prepare global forecasts in grid-point form of upper winds and upper air temperatures and to supply the forecasts to associated RAFCs. One of NCEP's main RAFC functions is to prepare and supply to users charts of forecast winds, temperatures, and significant weather.

For example, Figures 10-5 and 10-6, are originated by NCEP. The flight level of Figure 10-5 is 34,000 feet MSL, and Figure 10-6 is 45,000 feet MSL. This, along with the valid time of the chart and the data base time (data from which the forecast was derived), makes up the legend along an edge of each chart.

Forecast winds are expressed in knots for spot locations with directions and speed depicted in the same manner as the U.S. forecast winds and temperatures aloft chart (Figure 10-1). Forecast temperatures are depicted for spot locations inside circles that are expressed in degrees Celsius. For charts with flight levels (FL) at or below FL180 (18,000 feet), temperatures are depicted as negative (-) or positive (+). On charts for FLs above FL180, temperatures are always negative and no sign is depicted.

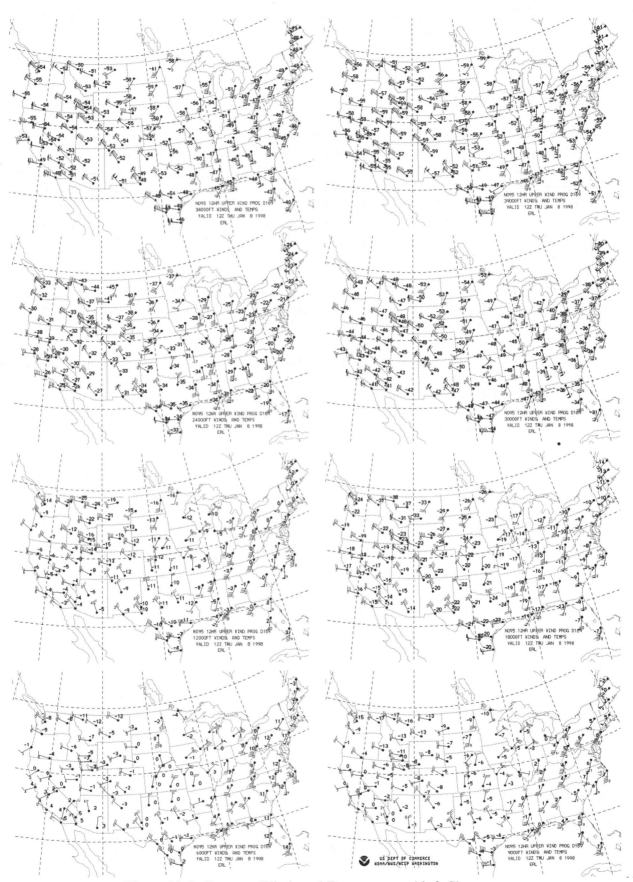

Figure 10-1. Forecast Winds and Temperatures Aloft Chart.

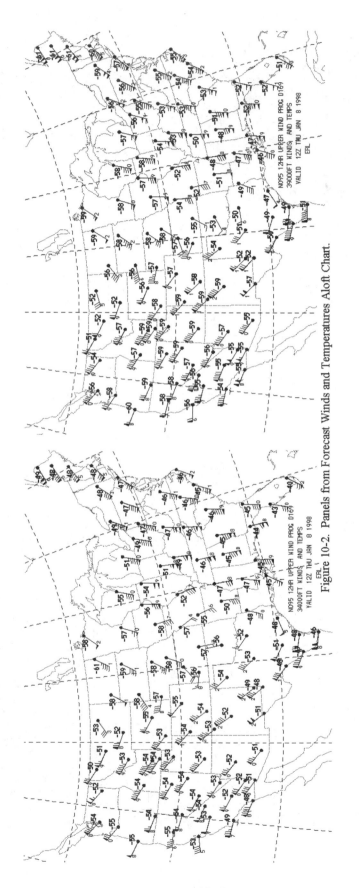

Figure 10-2. Panels from Forecast Winds and Temperatures Aloft Chart.

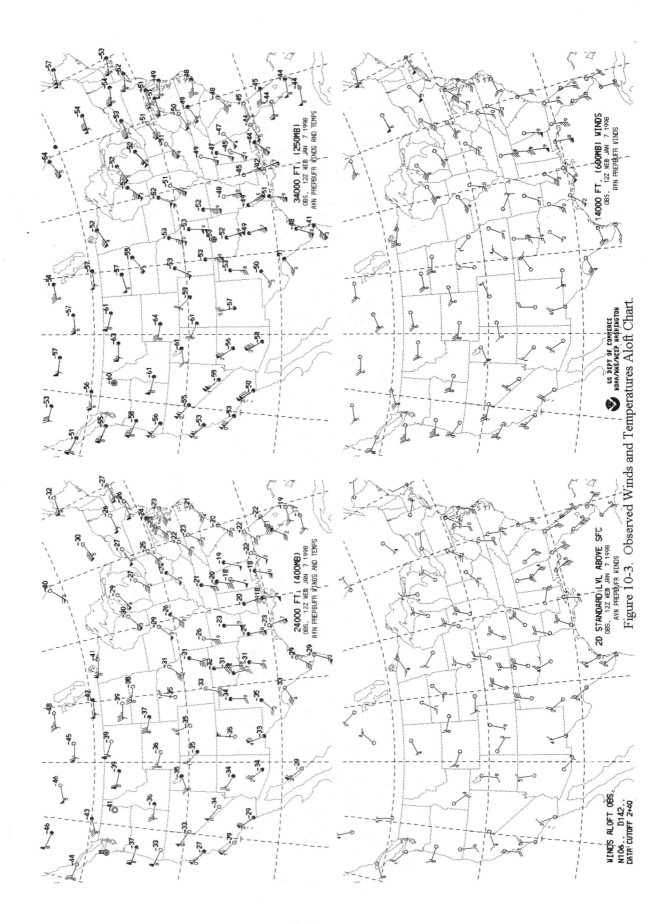

Figure 10-3. Observed Winds and Temperatures Aloft Chart.

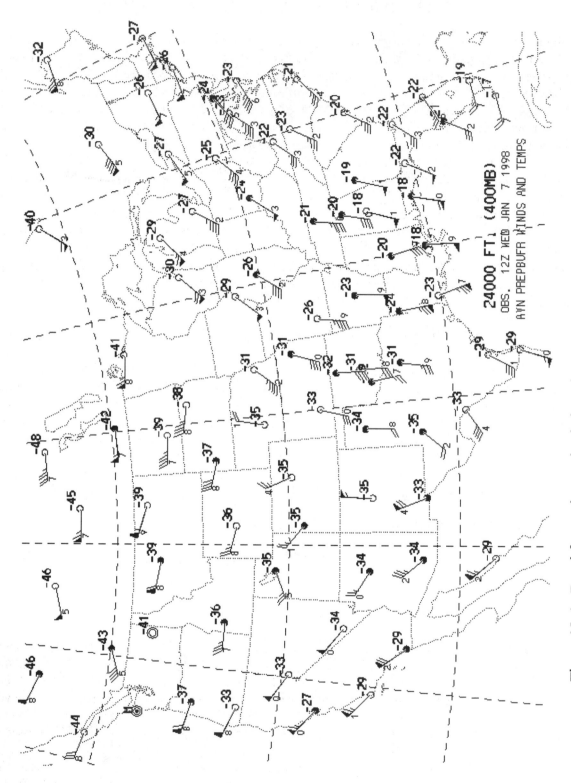

24000 FT₁ (400MB)
OBS. 12Z WED JAN 7 1998
AVN PREPBUFR WINDS AND TEMPS

Figure 10-4. Panel from Observed Winds and Temperatures Aloft Chart.

Section 11
SIGNIFICANT WEATHER PROGNOSTIC CHARTS

Significant weather prognostic charts (progs) (Figure 11-1) portray forecasts of selected weather conditions at specified valid times. Each valid time is the time at which the forecast conditions are expected to occur. Forecasts are made from a comprehensive set of observed weather conditions. The observed conditions are extended forward in time and become forecasts by considering atmospheric and environmental processes. Forecasts are made for various periods of time. A 12-hour prog is a forecast of conditions which has a valid time 12 hours after the observed data base time, thus a 12-hour forecast. A 24-hour prog is a 24-hour forecast, and so on. For example, a 12-hour forecast based on 00Z observations is valid at 12Z. Altitude information on the prog charts is referenced to mean sea level (MSL) and compatible with aviation. Altitudes below 18,000 feet are true altitudes while above 18,000 feet are pressure altitudes or flight levels (FL). The prog charts for the conterminous United States are generated for two general time periods. Day 1 progs are forecasts for the first 24-hour period. Day 2 progs are forecasts for the second 24-hour period. Day 1 prog charts are prepared for two altitude references in the atmosphere. Forecast information for the surface to 24,000 feet is provided by the low-level significant weather prog chart. Forecast information from above 24,000 to 60,000 feet is provided by the high-level significant weather prog chart. The day 2 prog chart is prepared without regard to altitude and is provided by the 36- and 48-hour surface prog chart.

U.S. LOW-LEVEL SIGNIFICANT WEATHER (SIG WX) PROG

The low-level significant weather prog chart (Figure 11-1) is a day 1 forecast of significant weather for the conterminous United States. Weather information provided pertains to the layer from surface to FL240 (400 mbs.) The information is provided for two forecast periods, 12 hours and 24 hours. The chart is composed of four panels. The two lower panels depict the 12- and 24-hour surface progs that are produced at Hydrometeorolgical Prediction Center (HPC) in Camp Springs, Maryland. The two upper panels depict the 12- and 24-hour significant weather progs that are produced at the Aviation Weather Center (AWC) in Kansas City, Missouri. The chart is issued four times a day; and the observation data base times for each issuance are 00Z, 06Z, 12Z, and 18Z.

SURFACE PROG PANELS

The surface prog panels display forecast positions and characteristics of pressure systems, fronts, and precipitation.

Surface Pressure Systems

Surface pressure systems are depicted by pressure centers, troughs, and, on selected panels, isobars. High and low pressure centers are identified by "Hs" and "Ls" respectively. The central pressure of each center is specified. Pressure troughs are identified by long dashed lines and labeled "TROF." Isobars are shown on selected panels. Isobars are drawn as solid lines and portray pressure patterns. The value of each isobar is identified by a two-digit code placed on each isobar. Isobars are drawn with intervals of 8 mbs relative to the 1,000 mb isobar. Note that this interval is larger than the 4-mb interval used on the surface analysis chart. The 8-mb interval provides a less sensitive analysis of pressure patterns than the 4-mb interval. Occasionally, nonstandard isobars will be drawn using 4-mb intervals to highlight patterns with weak pressure gradients. Nonstandard isobars are drawn as dashed lines. Examples of standard isobars drawn are the 992; 1,000; and 1,008 mb isobars.

11-2

Fronts

Surface fronts are depicted on each panel. Formats used are the standard symbols and three-digit characterization code used on the surface analysis chart. (See Section 5.)

Precipitation

Solid lines enclose precipitation areas. Symbols specify the forms and types of precipitation. (See Table 11-1.) A mix of precipitation is indicated by the use of two pertinent symbols separated by a slash. Identifying symbols are positioned within or adjacent to the precipitation areas. Precipitation conditions are described further by the use of shading. Stable precipitation events are classified as continuous or intermittent. Continuous precipitation is a dominant and widespread event and, therefore, shaded. Intermittent precipitation is a periodic and patchy event and unshaded. Shading is also used to characterize the coverage of unstable precipitation events. Areas with more than half coverage are shaded, and half or less coverage are unshaded. (See Table 11-2.) A bold dashed line is used to separate precipitation with contrasting characteristics within an outlined area. For example, a dashed line would be used to separate an area of snow from an area of rain.

Table 11-1 Standard Weather Symbols

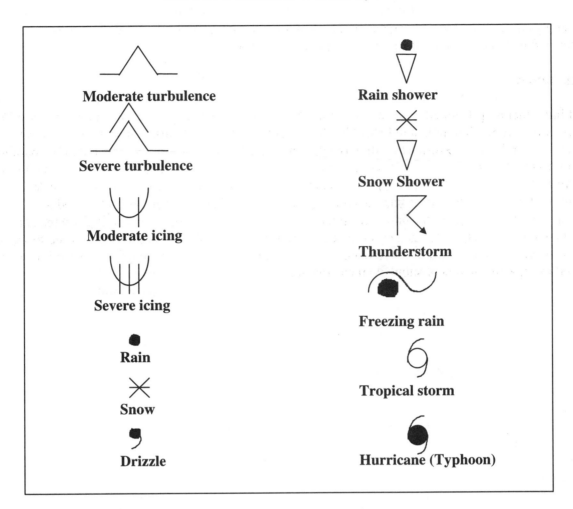

Table 11-2 Significant Weather Prog Symbols

Intermittent snow

Rain showers covering half or less the area

Continuous rain

Rain showers and thunderstorms covering more than half the area

SIGNIFICANT WEATHER PANELS

The significant weather panels display forecast weather flying categories, freezing levels, and turbulence for the layer surface to FL240. A legend on the chart illustrates symbols and criteria used for these conditions. (See Figure 11-1.)

Weather Flying Categories

The weather flying categories are visual flight rules (VFR), marginal VFR (MVFR), and instrument flight rules (IFR). Ceiling and visibility criteria used for each category are the same as used for the weather depiction chart. (See Section 6.) IFR areas are enclosed by solid lines. MVFR areas are enclosed by scalloped lines. All other areas are VFR.

Freezing Levels

The surface freezing level is depicted by a zigzag line and labeled "SFC." The surface freezing level separates above-freezing from below-freezing temperatures at the Earth's surface. Freezing levels aloft are depicted by thin, short dashed lines. Lines are drawn at 4,000-foot intervals beginning at 4,000 feet and labeled in hundreds of feet. For example, "80" identifies the 8,000-foot contour. Freezing level heights are referenced to MSL. The lines are discontinued where they intersect corresponding altitudes of the Rocky Mountains. The freezing level values for locations between lines is determined by linear interpolation. For example, the freezing level midway between the 4,000 and 8,000 foot lines is 6,000 feet. Areas with forecast multiple freezing levels have lines drawn to the highest freezing level. For example, with freezing levels forecast at 2,000, 6,000, and 8,000 feet, the analysis is drawn to the 8,000 foot value. Notice that not all freezing levels are identified with a multiple freezing level event. Information about the 2,000- and 6,000-foot freezing levels in this example would not be displayed. Surface-based multiple freezing levels are located over areas which have below-freezing temperatures at the surface and above-freezing temperatures within at least one layer aloft. Freezing rain and freezing drizzle (freezing precipitation) are associated with surface-based multiple freezing levels. The intersection of the surface freezing level line and freezing level contours encloses an area with surface-based multiple freezing levels.

Turbulence

Areas of moderate or greater turbulence are enclosed by bold, long dashed lines. Turbulence intensities are identified by symbols. The vertical extent of turbulence layers is specified by top and base heights in hundreds of feet. Height values are relative to MSL with the top and base heights separated by a line. A top height of "240" indicates turbulence at or above 24,000 feet. (The upper limit of the prog is 24,000 feet.) The base height is omitted where turbulence reaches the surface. For example, "080/ " identifies a turbulence layer from the surface to 8,000 feet MSL. Thunderstorms always imply a variety of hazardous conditions to aviation including moderate or greater turbulence. Generally, turbulence conditions implied with thunderstorms is not depicted on the chart. However, for added emphasis, moderate to severe turbulence surface to above 24,000 feet is depicted for areas that have thunderstorms with more than half coverage on the surface prog. Intensity symbols and layer altitudes appear within or adjacent to the forecast area.

USING THE CHART

The low-level significant weather prog chart provides an overview of selected flying weather conditions up to 24,000 feet for day 1. Much insight can be gained by evaluating the individual fields of pressure patterns, fronts, precipitation, weather flying categories, freezing levels, and turbulence displayed on the chart. In addition, certain inferences can be made from the chart. Surface winds can be inferred from surface pressure patterns. Structural icing can be inferred in areas which have clouds and precipitation, above freezing levels, and in areas of freezing precipitation. The low-level prog chart can also be used to obtain an overview of the progression of weather during day 1. The progression of weather is the change in position, size, and intensity of weather with time. Progression analysis is accomplished by comparing charts of observed conditions with the 12- and 24-hour prog panels. Progression analysis adds insight to the time-continuity of the weather from before flight time to after flight time. The low-level prog chart makes the comprehension of weather details easier and more meaningful. A comprehensive overview of weather conditions does not provide sufficient information for flight planning. Additional weather details are required. Essential weather details are provided by observed reports, forecast products, and weather advisories. Weather details are often numerous. An effective overview of observed and prog charts allows the many essential details to fit into place and have continuity.

36- AND 48-HOUR SURFACE PROG

The 36- and 48-hour surface prog chart (Figure 11-2) is a day 2 forecast of general weather for the conterminous United States. The chart is an extension of the day 1 U.S. low-level significant weather prog chart issued from the same observed data base time. These two prog charts make up a forecast package. The chart is issued twice daily. The observation data base times for each issuance are 00Z and 12Z. For example, a chart issued based on 00Z Tuesday observations has a 36-hour valid time of 12Z Wednesday and a 48-hour valid time of 00Z Thursday. The chart is composed of two panels and a forecast discussion. The two panels contain the 36- and 48-hour surface progs.

SURFACE PROG PANELS

The surface prog panels display forecast positions and characteristics of pressure patterns, fronts, and precipitation.

Surface Pressure Systems

Surface pressure systems are depicted by pressure centers, troughs, and isobars. Formats used for each feature are the same as used for the surface prog panels of the U.S. low-level significant weather prog chart.

Fronts

Surface fronts are depicted by using the standard symbols and three-digit characterization code used on the surface analysis chart. (See Section 5.)

Precipitation

Precipitation areas are outlined on each panel. Formats used to locate and characterize precipitation are the same as used for the surface prog panels of the U.S. low-level prog chart.

FORECAST DISCUSSION

The forecast discussion is a discussion of the day 1 and day 2 forecast package. The discussion will include identification and characterization of weather systems and associated weather conditions portrayed on the prog charts.

USING THE CHART

The 36- and 48-hour surface prog chart provides an outlook of general weather conditions for day 2. The 36- and 48-hour prog can also be used to assess the progression of weather through day 2.

HIGH-LEVEL SIGNIFICANT WEATHER PROG

The high-level significant weather prog (Figures 11-3 and 11-4) is a day 1 forecast of significant weather. Weather information provided pertains to the layer from above 24,000 to 60,000 feet (FL250-FL600). The prog covers a large portion of the Northern Hemisphere and a limited portion of the Southern Hemisphere. Coverage ranges from the eastern Asiatic coast eastward across the Pacific, North America, and the Atlantic into Europe and northwestern Africa. The prog extends southward into northern South America. The area covered by the prog is divided into sections. Each section covers a part of the forecast area. Some sections overlap. The various sections are formatted on polar or Mercator projection background maps and issued as charts. Each prog chart is issued four times a day. The valid times are 00Z, 06Z, 12Z, and 18Z. Conditions routinely appearing on the chart are jet streams, cumulonimbus clouds, turbulence, and tropopause heights. Surface fronts are also included to add perspective. Other conditions will appear on the chart as pertinent. They are tropical cyclones, squall lines, volcanic eruption sites, and sandstorms and dust storms.

Jet Streams

Jet streams with a maximum speed of more than 80 knots are identified by bold lines. Jet stream lines lie along the core of maximum winds. Arrowheads on the lines indicate the orientation of each jet stream. Double hatched lines positioned along the jet core identify changes of wind speed. These speed indicators are drawn at 20-knot intervals and begin with 100 knots. Wind speed maximums along the jet core are characterized by wind symbols and altitudes. A standard wind symbol (shaft, pennants, and barbs) is placed at each pertinent position to identify velocity. The altitude in hundreds of feet prefaced with "FL" is placed adjacent to each wind symbol.

Example:

Cumulonimbus Clouds

Cumulonimbus clouds (CBs) are thunderstorm clouds. Areas of CBs meeting select criteria are enclosed by scalloped lines. The criteria are widespread CBs within an area or along a line with little or no space between individual clouds, and CBs are embedded in cloud layers or concealed by haze or dust. The prog does not display isolated or scattered CBs (one-half or less coverage) which are not embedded in clouds, haze, or dust. Cumulonimbus areas are identified with "CB" and characterized by coverage and tops. Coverages are identified as isolated (ISOL), occasional (OCNL), and frequent (FRQ). Isolated and occasional CBs are further characterized as embedded (EMBD.) Coverage values for the identifiers are: isolated - less than 1/8; occasional - 1/8 to 4/8; and frequent - more than 4/8. Tops are identified in hundreds of feet using the standard top and base format. Bases extend below 24,000 feet (below the prog's forecast layer) and are encoded "XXX." The identification and characterization of each cumulonimbus area will appear within or adjacent to the outlined area. Thunderstorms always imply a variety of aviation hazards including moderate or greater turbulence and hail.

Examples:

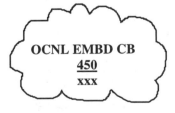

Turbulence

Areas of moderate or greater turbulence are enclosed by bold dashed lines. Turbulence conditions identified are those associated with wind shear zones and mountain waves. Wind shear zones include speed shears associated with jet streams and areas with sharply curved flow. Turbulence associated with thunderstorms is not identified. (Thunderstorms imply turbulence.) Turbulence intensities are identified by symbols. The vertical extent of turbulence layers is specified by top and base heights in hundreds of feet. Turbulence bases which extends below the layer of the chart are identified with "XXX." Top and base heights are separated by a line. Height values are pressure altitudes. For example, "310/XXX" identifies a layer of turbulence from below FL240 to FL310.

Example:

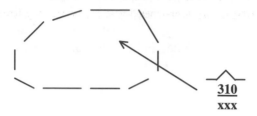

Tropopause Heights

Tropopause heights are plotted in hundreds of feet at selected locations. Heights are enclosed by rectangles. Centers of high and low heights are identified with "H" and "L" respectively along with their heights and enclosed by polygons.

Examples:

Surface Fronts

Surface fronts are depicted on the prog to provide added perspective. Symbols used are the standard symbols used on the surface analysis chart. Movements of fronts are identified at selected positions. A vector with a number plotted adjacent to the vector identifies the direction and speed of movement. (See Section 5.)

Example:

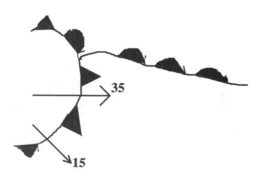

Tropical Cyclones

The positions of hurricanes, typhoons, and tropical storms are depicted by symbols. The only difference between the hurricane/typhoon symbol and tropical storm symbol is the circle of the hurricane/typhoon symbol is shaded in. When pertinent, the name of each storm is positioned adjacent to the symbol. Cumulonimbus cloud activity meeting chart criteria is identified and characterized relative to each storm.

Example:

 **Tropical Storm**

Squall Lines

Severe squall lines are lines of CBs with 5/8 coverage or greater. Squall lines are identified by long dashed lines, and each dash is separated by a "v." Cumulonimbus cloud activity meeting chart criteria is identified and characterized with each squall line.

Example:

———— V ———— V ————

Volcanic Eruption Sites

Volcanic eruption sites are identified by a trapezoidal symbol. The dot on the base of the trapezoid locates the latitude and longitude of the volcano. The name of the volcano, its latitude, and its longitude are noted adjacent to the symbol. Pertinent SIGMETs containing information regarding volcanic ash will be in effect.

Example:

TUNGURAHUA
1.5S   78.4W

Sandstorms and Dust Storms

Areas of widespread sandstorms and dust storms are labeled by symbol. The symbol with the arrow depicts areas of widespread sandstorm or dust storm, while the symbol without the arrow depicts severe sandstorm or dust haze.

Example:

USING THE CHART

The high-level sig weather prog is used to get an overview of selected flying weather conditions above 24,000 feet. Much insight can be gained by evaluating jet streams, cumulonimbus clouds, turbulence, associated surface fronts, significant tropical storm complexes including tropical cyclones, squall lines, sandstorms, and dust storms.

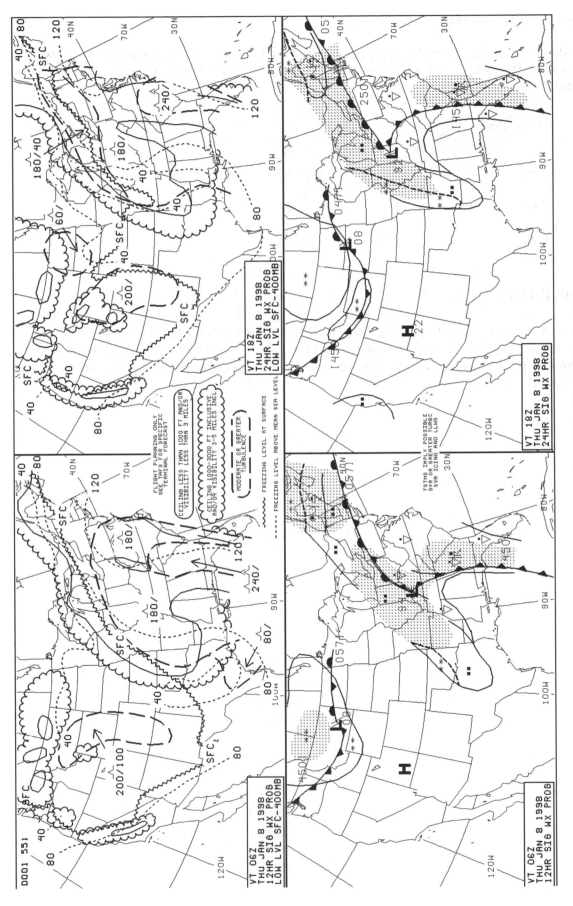

Figure 11-1. U.S. Low-Level Significant Weather Prog.

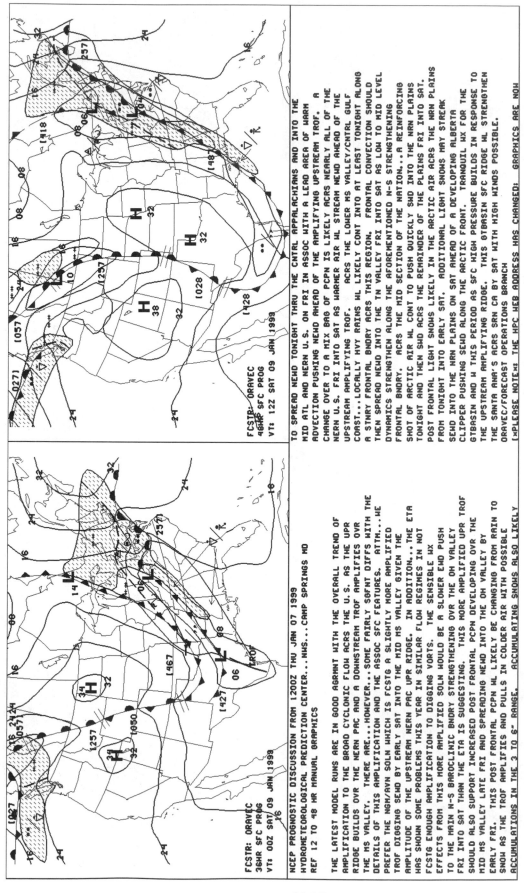

Figure 11-2. U.S. Low-Level 36- and 48-hour Significant Weather Prog.

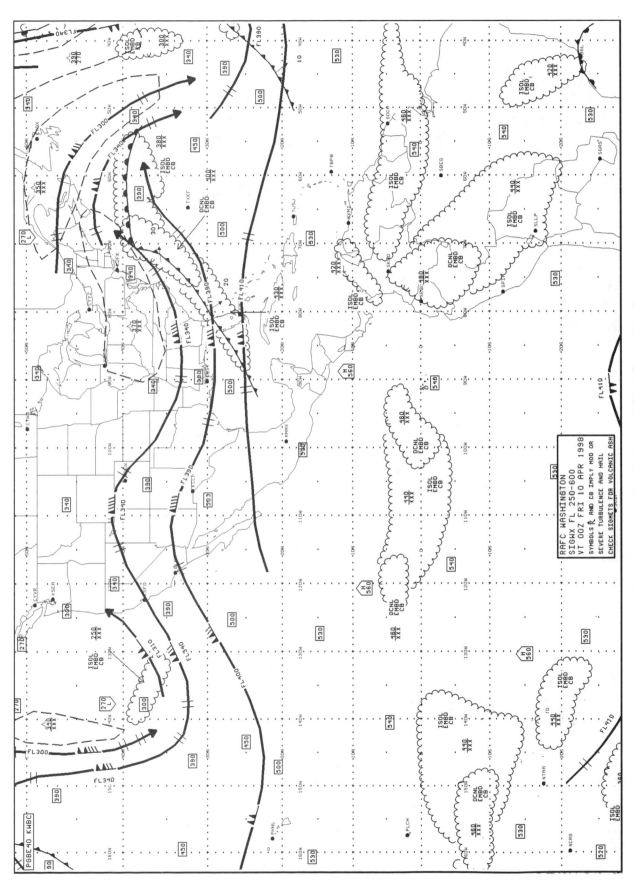

Figure 11-3. U.S. High-Level Significant Weather Prog.

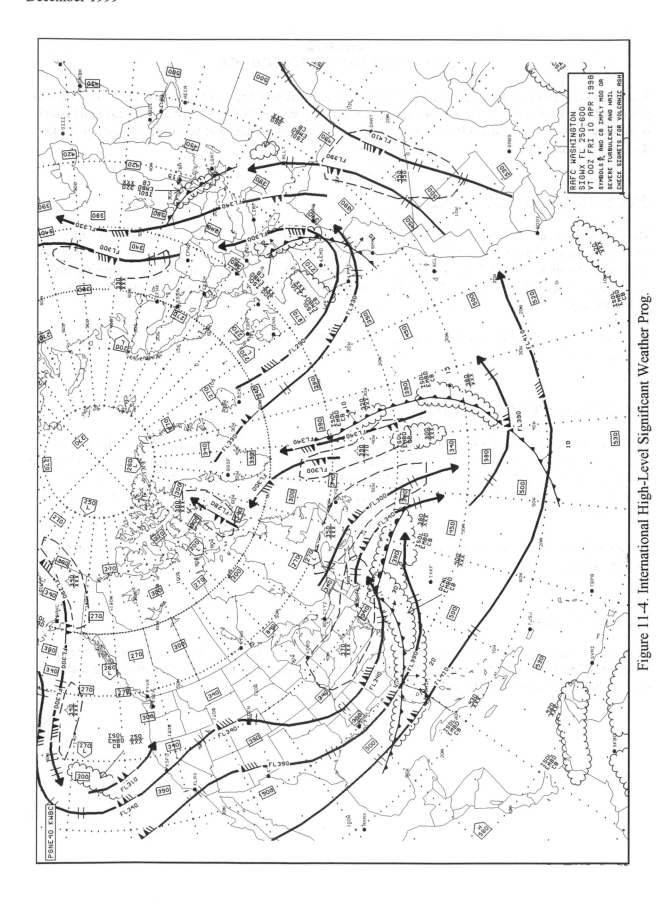

Figure 11-4. International High-Level Significant Weather Prog.

Section 12
CONVECTIVE OUTLOOK CHART

The convective outlook chart (Figure 12-1) delineates areas forecast to have thunderstorms. This chart is presented in two panels. The left-hand panel is the Day 1 Convective Outlook, and the right-hand panel is the Day 2 Convective Outlook. These guidance products are produced at the Storm Prediction Center (SPC) in Norman, OK.

DAY 1 CONVECTIVE OUTLOOK

The Day 1 Convective Outlook (Figure 12-1) outlines areas in the continental United States where thunderstorms are forecasted during the Day 1 period. It is issued five times daily. The first issuance is 06Z and is the initial Day 1 Convective Outlook that is valid 12Z that day until 12Z the following day. The other issuances are 1300Z, 1630Z, 2000Z, and 0100Z, and all issuances are valid until 12Z the following day.

The outlook issued qualifies the level of risk (i.e., SLGT, MDT, HIGH) as well as the areas of general thunderstorms.

DAY 2 CONVECTIVE OUTLOOK

The Day 2 Convective Outlook contains the same information as the Day 1 Convective Outlook. It is issued twice a day. It is initially issued at 0830Z during standard time and 0730Z during daylight time. It is updated at 1730Z. The timeframe covered is from 12Z the following day to 12Z the next day. For example, if today is Monday, the Day 2 Convective Outlook will cover the period 12Z Tuesday to 12Z Wednesday.

The outlook issued qualifies the level of risk (i.e., SLGT, MDT, HIGH) as well as the areas of general thunderstorms.

LEVELS OF RISK

Risk areas come in three varieties and are based on the expected number of severe thunderstorm reports per geographical unit and forecaster confidence. Table 12-1 indicates the labels that appear on both the Day 1 and Day 2 Convective Outlook charts.

Table 12-1 Notations of Risk

NOTATION	EXPLANATION
SEE TEXT	Used for those situations where a SLGT risk was considered but at the time of the forecast, was not warranted.
SLGT (Slight risk)	A high probability of 5 to 29 reports of 1 inch or larger hail, and/or 3-5 tornadoes, and/or 5 to 29 wind events,... or... a low/moderate probability of moderate to high risk being issued later if some conditions come together
MDT (Moderate risk)	A high probability of at least 30 reports of hail 1 inch or larger; or 6-19 tornadoes; or numerous wind events (30).
HIGH (High risk)	A high probability of at least 20 tornadoes with at least two of them rated F3 (or higher), or an extreme derecho causing widespread (50 or more) wind events with numerous higher-end wind (80 mph or higher) and structural damage reports

SEE TEXT is used for those situations where a slight risk was considered, but at the time of the forecast, was not warranted. Although there is no severe outlook for the labeled area, users should read the text of the convective outlook (AC) forecast message to learn more about the potential for a threat to develop if some particular conditions do come together.

Slight (SLGT) risk implies well-organized severe thunderstorms are expected but in small numbers and/or low coverage.

Moderate (MDT) risks imply a greater concentration of severe thunderstorms, and in most situations, greater magnitude of severe weather.

High (HIGH) risk almost always means a major severe weather outbreak is expected, with great coverage of severe weather and enhanced likelihood of extreme severe events (i.e., violent tornadoes or unusually intense damaging wind events). SPC issues a public information statement (PWO) describing a "particularly dangerous situation" when HIGH risk areas are in effect, and it sometimes issues a PWO for MDT risk situations. Some National Weather Service (NWS) offices will include in their public forecasts the phrase "some thunderstorms may be severe" when a MDT or HIGH risk is issued.

In addition to the severe risk areas, general thunderstorms (non-severe) are outlined, but with no label on the graphic map.

USING THE CHART

The Day 1 and Day 2 Convective Outlooks Charts are flight planning tools used to determine forecast areas of thunderstorms.

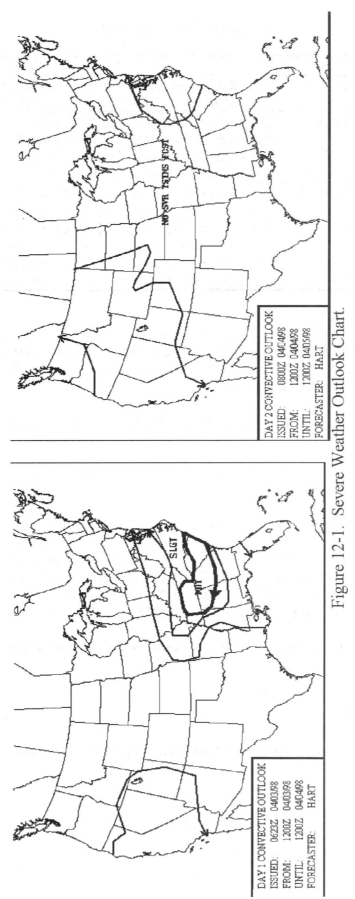

Figure 12-1. Severe Weather Outlook Chart.

Section 13
VOLCANIC ASH ADVISORY CENTER (VAAC) PRODUCTS

The Volcanic Ash Advisory Center (VAAC) may issue two products when there is a volcanic eruption: the Volcanic Ash Advisory Statement (VAAS) and forecast charts of ash dispersion. The U.S. VAACs are the AAWU in Anchorage, Alaska, and the Washington, D.C. VAAC located in Camp Springs, Maryland. Other international centers contribute to the tracking of volcanic ash events. The VAACs do not issue routine products but create and issue them when a volcanic eruption occurs. The products are based on information from PIREPs, MWO SIGMETs, satellite observations, and volcanic observatory reports. Since the products are triggered by the occurrence of an eruption, pilot reports concerning volcanic activity are extremely important.

VOLCANIC ASH ADVISORY STATEMENT (VAAS)

Usually the first VAAC product to be issued is the Volcanic Ash Advisory Statement(VAAS). The VAAS is required to be issued within 6 hours of an eruption and every 6 hours after that. However, it can be issued more frequently if new information about the eruption is received. The VAAS summarizes the currently known information about the eruption. It may include the location of the volcano, height of the volcano summit, height of the ash plume, a latitude/longitude box of the ash dispersion cloud, and a forecast of ash dispersion. The height of the ash cloud is estimated by meteorologists analyzing satellite imagery and satellite cloud drift winds combined with any pilot reports, volcano observatory reports, and upper-air wind reports. The VAASs are transmitted to users via the Global Telecommunications System (GTS), the World Area Forecast System (WAFS), the Aeronautical Fixed Telecommunications Network (AFTN), the FAA communications system (WMWCR), and the NWS Family of Services. In addition, VAASs are available on several Internet sites listed on the last page of this document.

Example of a VAAS:
FVAK20 PANC 190323
VOLCANIC ASH ADVISORY - ALERT
ALASKA AVIATION WEATHER UNIT
NATIONAL WEATHER SERVICE ANCHORAGE AK
ISSUED 0300 UTC SUNDAY JULY 19 1998 BY ANCHORAGE VAAC

VOLCANO: KARYMSKY (1000-13) 98-01
KAMCHATKA 54.05N 159.43E 1486 M 4875 FT

SOURCES OF INFORMATION: PILOT REPORT

ERUPTION DETAILS: ERUPTION TO FL100 REPORTED BY PILOT REPORT AT
19/0200 UTC VIA WASHINGTON DC VAAC.

ASH CLOUD DESCRIPTION: N/A

ASH CLOUD TRAJECTORY: NE 10 KT.

12 HOUR OUTLOOK: IF ASH PERSISTS ALOFT AT 12 HOURS THE FORECAST AREA
FROM THE PUFF MODEL BELOW 15000FT IS 56N 161E, 55N 166E, 54N 165E, 55N
162E.

ADDITIONAL INFORMATION: NO ERUPTION VISIBLE ON SATELLITE IMAGERY
DUE TO CLOUD IN AREA.

THIS WILL BE THE ONLY ADVISORY ISSUED FOR THIS EVENT.

DAC JUL98 AAWU

VOLCANIC ASH FORECAST TRANSPORT AND DISPERSION (VAFTAD) CHART

The Volcanic Ash Forecast Transport and Dispersion (VAFTAD) Chart, Figures 13-1 and 13-2, is
generated by a three-dimensional time-dependent dispersion model developed by the National Oceanic
and Atmospheric Administration (NOAA) Air Resources Laboratory (ARL). The VAFTAD model
focuses on hazards to aircraft flight operations caused by a volcanic eruption with an emphasis on the
ash cloud location in time and space. It uses National Centers for Environmental Prediction (NCEP)
forecast data to determine the location of ash concentrations over 6-hour and 12-hour intervals, with
valid times beginning 6, 12, 24, and 36 hours following a volcanic eruption. This computer-prepared
chart is not issued on a routine basis, but only as volcanic eruptions are reported. Since the VAFTAD
chart is triggered by the occurrence of volcanic eruption, PIREPs concerning volcanic activity are very
important. Initial input to the VAFTAD model run and the resulting chart include: geographic region,
volcano name, volcano latitude and longitude, eruption date and time, and initial ash cloud height.
Utilizing the NCEP meteorological forecast guidance, volcanic ash particle transport and dispersion are
depicted horizontally and vertically through representative atmospheric layers. The charts from an
actual eruption will be labeled with ALERT. Another possible reason to generate a chart could be for
potential volcanic eruption. This chart would be labeled WATCH as shown on Figure 13-1.

VAFTAD PRODUCT

The VAFTAD product presents the relative concentrations of ash following a volcanic eruption for three layers of the atmosphere in addition to a composite of ash concentration through the atmosphere. Atmospheric layers depicted are: surface to flight level (FL) 200, surface to FL550 (composite), FL200 to FL350, and FL350 to FL550. Figure 13-1 shows 8 panels of ash cloud relative concentrations for 12 to 24 hours; and Figure 13-2 shows 18 to 24 hours after a volcanic eruption. Note that the first 6 hours after the volcanic eruption are not depicted. An appropriate SIGMET will be issued by an MWO for that period concerning the volcanic eruption and the area affected by the ash cloud. The four panels in any column are valid for the same time interval (specified and located below the third panel). The top three panels in each column provide the ash location and relative concentrations for an atmospheric layer, identified by top and bottom flight levels. The highest layer is at the top of the chart. Volcano eruption information is given in the legend at the lower left (see Figure 13-1) which includes the volcano name (with location symbol), latitude and longitude, eruption date and time, duration, and ash column height.

USING THE CHART

The VAFTAD chart is strictly for advanced flight planning purposes. It is not intended to take the place of SIGMETs regarding volcanic eruptions and ash.

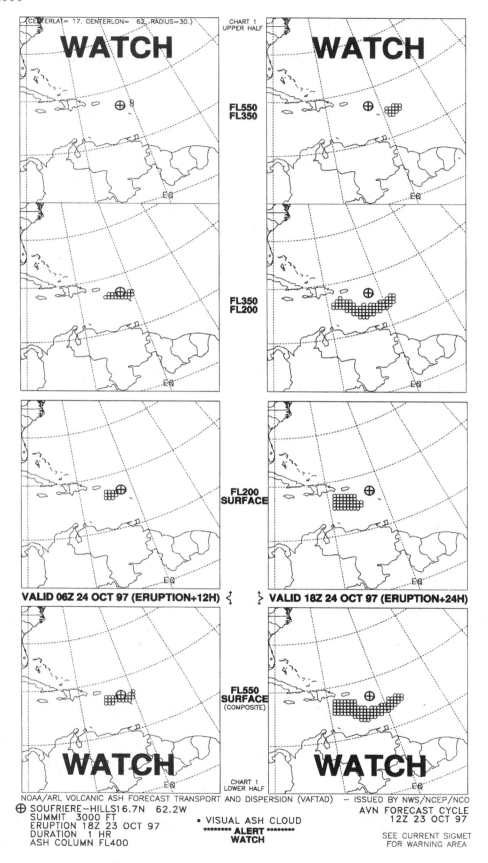

Figure 13-1. Volcanic Ash Forecast Chart.

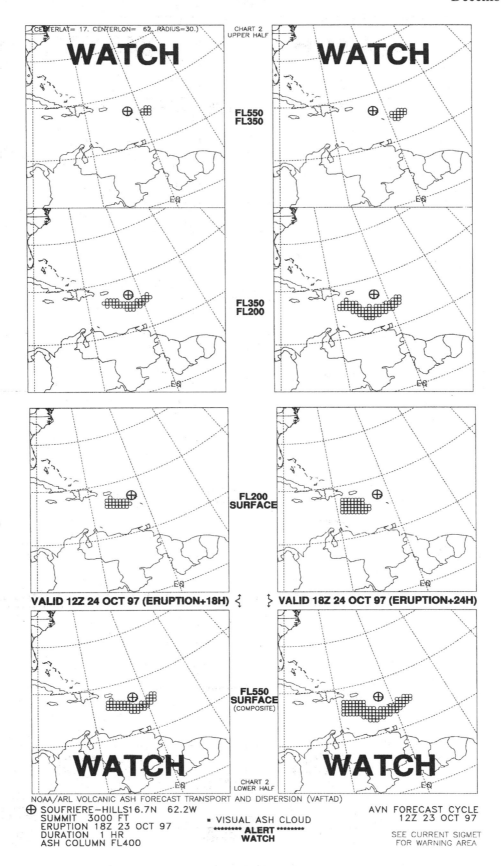

Figure 13-2. Volcanic Ash Forecast Chart.

Section 14
TURBULENCE LOCATIONS, CONVERSION AND DENSITY ALTITUDE TABLES, CONTRACTIONS AND ACRONYMS, SCHEDULE OF PRODUCTS, NATIONAL WEATHER SERVICE STATION IDENTIFIERS, WSR-88D SITES, AND INTERNET ADDRESSES

This section provides text, graphs, and tables that can be used by the pilot to further understand the weather. Information included covers:

1. Locations of probable turbulence
2. Standard conversions table
3. Density altitude and chart
4. Contractions and acronyms
5. Scheduled issuance and valid times of forecast products
6. National Weather Service station identifiers and WSR-88D sites
7. Internet addresses

LOCATIONS OF PROBABLE TURBULENCE

Turbulence occurs due to either terrain features or weather phenomenon which can produce intensities from light to extreme. The type and intensity of the turbulence will depend on the situations as described in the following paragraphs.

LIGHT TURBULENCE

Light turbulence can be caused by obstruction of the wind in hilly or mountainous terrain. Even with light winds, there can be enough displacement of the wind to produce small-scale eddies or turbulence.

Weather conditions that can cause light turbulence are associated with clear-air convective currents over a heated surface or near and in small cumulus clouds. Weak wind shear in the vicinity of troughs aloft, lows aloft, jet streams, or the tropopause can cause light turbulence. Also in the lower 5,000 feet of the atmosphere, light turbulence can occur when winds are near 15 knots or where the air is colder than the underlying surfaces.

MODERATE TURBULENCE

Moderate turbulence will be reported in mountainous areas with a wind component of 25 to 50 knots perpendicular to and near the level of the ridge. The turbulence will be located at all levels from the surface to 5,000 feet above the tropopause. The areas most likely to have moderate turbulence is within 5,000 feet of the ridge level, at the base of relatively stable layers below the base of the tropopause, or within the tropopause layer. The turbulence will extend downstream from the lee of the ridge for 150 to 300 miles.

Also, moderate turbulence can be encountered in and near towering cumuliform clouds and thunderstorms (in the dissipating stage).

Moderate turbulence can occur in the lower 5,000 feet of the troposphere when surface winds are 30 knots or more, where heating of the underlying surface is unusually strong, where there is an invasion of very cold air, or in fronts aloft.

Wind shear in the vertical direction that exceeds 6 knots per 1,000 feet and/or horizontal wind shear that exceeds 18 knots per 150 miles will produce moderate turbulence.

SEVERE TURBULENCE

Severe turbulence is likely in mountainous areas with a wind component exceeding 50 knots perpendicular to and near the level of the ridge. The location of the severe turbulence will be in 5,000-foot layers at and below the ridge level in rotor clouds or rotor action, at the tropopause, and sometimes at the base of other stable layers below the tropopause. The severe turbulence will extend downstream from the lee of the ridge for 50 to 150 miles.

Severe turbulence can be encountered in and near growing and mature thunderstorms and occasionally in other towering cumuliform clouds.

Severe turbulence will also occur 50 to 100 miles on the cold side of the center of the jet stream, in troughs aloft, and in lows aloft where vertical wind shear exceeds 10 knots per 1,000 feet, and horizontal wind shear exceeds 40 knots per 150 miles.

EXTREME TURBULENCE

Extreme turbulence will be found in mountain wave situations. The turbulence will be located in and below the level of well-developed rotor clouds. Sometimes the turbulence extends to the ground.

Besides mountain wave situations, extreme turbulence will occur in severe thunderstorms. A severe thunderstorm is indicated by large hailstones (diameter ¾ inch or greater), strong radar echoes, or continuous lightning.

STANDARD CONVERSION TABLE

This table can be used as a quick reference for conversion between metric and English units.

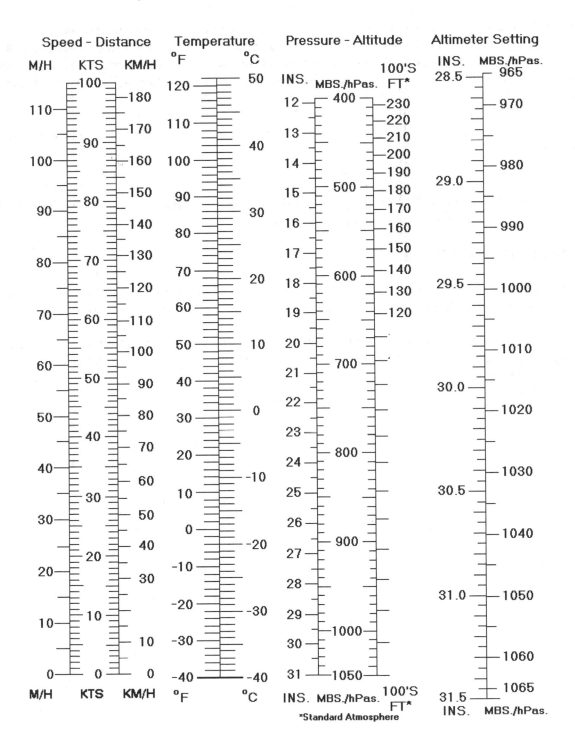

Figure 14-1. Standard Conversion Table.

DENSITY ALTITUDE

Density altitude can affect the takeoff, climb, and landing performance of any aircraft. The distance required to take off and land and the rate of climb are affected by density altitude.

Aircraft will perform better in low density altitude conditions. Low density altitude conditions exist when the air is dense. This occurs when the temperature is cold combined with a high pressure system. The air is the most dense in this situation and the aircraft will perform as if it were at a lower altitude. For example, a plane is at an airport with a station elevation of 7,000 feet MSL. The atmospheric conditions at that airport indicate a low density altitude situation. The density altitude is calculated to be 5,500 feet MSL. The plane will perform as if it were at 5,500 feet MSL instead of 7,000 feet MSL. This low density altitude situation will decrease takeoff and landing roll while increasing the initial rate of climb.

While low density altitude increases aircraft performance, high density altitude can lead to an aircraft accident. High density altitude situations are more prevalent at higher elevations. High temperatures combined with a low pressure system will produce a high density altitude situation. (The air is least dense in this situation.) Airports in mountainous terrain are more susceptible to high density altitude situations because they already have a high station elevation. The combination of a high station elevation, high temperatures, and low pressure will produce a very high density altitude situation. For example, a plane is at an airport with a station elevation of 7,000 feet MSL. Using the values of station elevation, temperature, and pressure, the density altitude is calculated to be 12,000 feet MSL. Any aircraft taking off or landing at that airport will perform as if it were at an airport with a station elevation of 12,000 MSL. For some aircraft, a high density altitude situation will indicate an altitude higher than the service ceiling of that specific aircraft. In that case, if a pilot attempts to take off during a high density situation, the aircraft will not be able to gain altitude but stay in ground effect and possibly crash.

Use Figure 14-2 to find density altitude either on the ground or aloft. Set the aircraft's altimeter at 29.92 inches. The altimeter will indicate pressure altitude. Read the outside air temperature. Enter the graph at the pressure altitude value and move horizontally to the temperature value. Read the density altitude from the sloping lines.

Examples:
Density altitude in flight: Pressure altitude is 9,500 feet and the temperature is -8 degrees C. Find 9,500 feet on the left of the chart and move to -8 degrees C. Density altitude is 9,000 feet. See dot on the chart that is labeled number 1.

Density altitude for takeoff: Pressure altitude is 4,950 feet and the temperature is 97 degrees F. Enter the graph at 4,950 feet and move across to 97 degrees F. Density altitude is 8,200 feet. See dot on the chart that is labeled number 2.

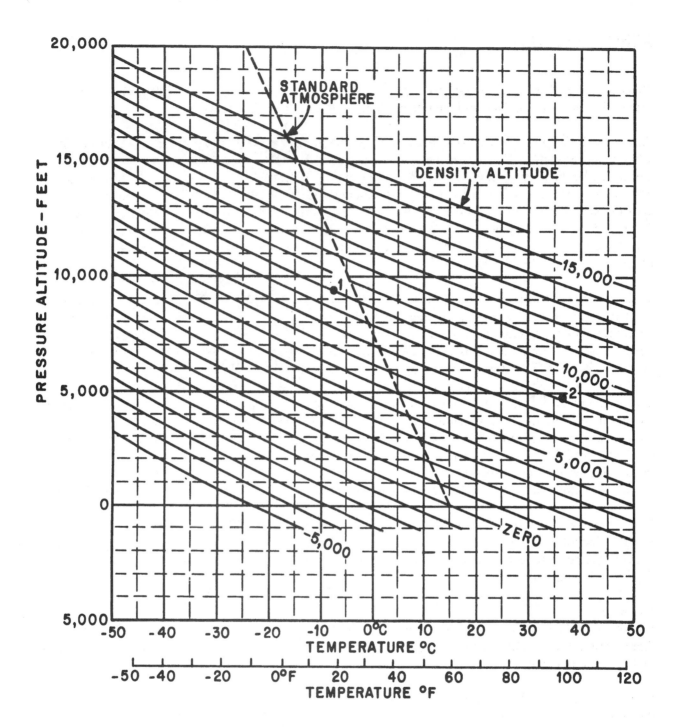

Figure 14-2. Density Altitude Computation Chart.

CONTRACTIONS AND ACRONYMS

Contractions and acronyms are used extensively in surface reports, pilot reports, and forecasts.

A

AAWU – Alaskan Aviation Weather Unit
ABNDT - Abundant
ABNML - Abnormal
ABT - About
ABV - Above
AC - Convective outlook or altocumulus
ACC - Altocumulus castellanus
ACCUM - Accumulate
ACFT - Aircraft
ACLT - Accelerate
ACLTD - Accelerated
ACLTG - Accelerating
ACLTS - Accelerates
ACPY - Accompany
ACRS - Across
ACSL - Altocumulus standing lenticular
ACTV - Active
ACTVTY - Activity
ACYC - Anticyclone
ADJ - Adjacent
ADL - Additional
ADQT - Adequate
ADQTLY - Adequately
ADRNDCK - Adirondack
ADVCT - Advect
ADVCTD - Advected
ADVCTG - Advecting
ADVCTN - Advection
ADVCTS - Advects
ADVN - Advance
ADVNG - Advancing
ADVY - Advisory
ADVYS - Advisories
AFCT - Affect
AFCTD - Affected
AFCTG - Affecting
AFDK - After dark
AFOS - Automated Field Operations System
AFSS - Automated Flight Service Station
AFT - After
AFTN - Afternoon
AGL - Above ground level
AGN - Again

AGRD - Agreed
AGRS - Agrees
AGRMT - Agreement
AHD - Ahead
AK - Alaska
AL - Alabama
ALF - Aloft
ALG - Along
ALGHNY - Allegheny
ALQDS - All quadrants
ALSTG - Altimeter setting
ALT - Altitude
ALTA - Alberta
ALTHO - Although
ALTM - Altimeter
ALUTN - Aleutian
AMD - Amend
AMDD - Amended
AMDG - Amending
AMDT - Amendment
AMP - Amplify
AMPG - Amplifying
AMPLTD - Amplitude
AMS - Air mass
AMT - Amount
ANLYS - Analysis
ANS - Answer
AOA - At or above
AOB - At or below
AP - Anomalous Propagation
APCH - Approach
APCHG - Approaching
APCHS - Approaches
APLCN - Appalachian
APLCNS - Appalachians
APPR - Appear
APPRG - Appearing
APPRS - Appears
APRNT - Apparent
APRNTLY - Apparently
APRX - Approximate
APRXLY - Approximately
AR - Arkansas
ARL – Air Resources Lab
ARND - Around
ARPT - Airport

ASAP - As soon as possible
ASSOCD - Associated
ASSOCN - Association
ATLC - Atlantic
ATTM - At this time
ATTN - Attention
AVBL - Available
AVG - Average
AVN - Aviation model
AWC – Aviation Weather Center
AWT - Awaiting
AZ - Arizona
AZM - Azimuth

B

BACLIN - Baroclinic
BAJA - Baja, California
BATROP - Barotropic
BC - British Columbia
BCH - Beach
BCKG - Backing
BCM - Become
BCMG - Becoming
BCMS - Becomes
BDA - Bermuda
BDRY - Boundary
BFDK - Before dark
BFR - Before
BGN - Begin
BGNG - Beginning
BGNS - Begins
BHND - Behind
BINOVC - Breaks in overcast
BKN - Broken
BLD - Build
BLDG - Building
BLDUP - Buildup
BLKHLS - Black Hills
BLKT - Blanket
BLKTG - Blanketing
BLKTS - Blankets
BLO - Below clouds
BLW - Below
BLZD - Blizzard
BN - Blowing sand
BND - Bound
BNDRY - Boundary
BNDRYS - Boundaries
BNTH - Beneath
BOOTHEEL - Bootheel

BR - Branch
BRF - Brief
BRK - Break
BRKG - Breaking
BRKHIC - Breaks in higher clouds
BRKS - Breaks
BRKSHR - Berkshire
BRM - Barometer
BS - Blowing snow
BTWN - Between
BYD - Beyond

C

C - Celsius
CA - California
CAA - Cold air advection
CARIB - Caribbean
CASCDS - Cascades
CB - Cumulonimbus
CC - Cirrocumulus
CCLDS - Clear of clouds
CCLKWS - Counterclockwise
CCSL - Cirrocumulus standing lenticular
CDFNT - Cold front
CFP - Cold front passage
CHC - Chance
CHCS - Chances
CHG - Change
CHGD - Changed
CHGG - Changing
CHGS - Changes
CHSPK - Chesapeake
CI - Cirrus
CIG - Ceiling
CIGS - Ceilings
CLD - Cloud
CLDNS - Cloudiness
CLDS - Clouds
CLKWS - Clockwise
CLR - Clear
CLRG - Clearing
CLRS - Clears
CMPLX - Complex
CNCL - Cancel
CNCLD - Canceled
CNCLG - Canceling
CNCLS - Cancels
CNDN - Canadian
CNTR - Center
CNTRD - Centered

CNTRL - Central
CNTY - County
CNTYS - Counties
CNVG - Converge
CNVGG - Converging
CNVGNC - Convergence
CNVTN - Convection
CNVTV - Convective
CNVTVLY - Convectively
CONFDC - Confidence
CO - Colorado
COMPR - Compare
COMPRG - Comparing
COMPRD - Compared
COMPRS - Compares
COND - Condition
CONT - Continue
CONTD - Continued
CONTLY - Continually
CONTG - Continuing
CONTRAILS - Condensation trails
CONTS - Continues
CONTDVD - Continental Divide
CONUS - Continental U.S.
COORD - Coordinate
COR - Correction
CPBL - Capable
CRC - Circle
CRLC - Circulate
CRLN - Circulation
CRNR - Corner
CRNRS - Corners
CRS - Course
CS - Cirrostratus
CSDR - Consider
CSDRBL - Considerable
CST - Coast
CSTL - Coastal
CT - Connecticut
CTGY - Category
CTSKLS - Catskills
CU - Cumulus
CUFRA - Cumulus fractus
CVR - Cover
CVRD - Covered
CVRG - Covering
CVRS - Covers
CWSU - Center Weather Service Units
CYC - Cyclonic
CYCLGN - Cyclogenesis

D

DABRK - Daybreak
DALGT - Daylight
DBL - Double
DC - District of Columbia
DCR - Decrease
DCRD - Decreased
DCRG - Decreasing
DCRGLY - Decreasingly
DCRS - Decreases
DE - Delaware
DEG - Degree
DEGS - Degrees
DELMARVA - Delaware-Maryland-Virginia
DFCLT - Difficult
DFCLTY - Difficulty
DFNT - Definite
DFNTLY - Definitely
DFRS - Differs
DFUS - Diffuse
DGNL - Diagonal
DGNLLY - Diagonally
DIGG - Digging
DIR - Direction
DISC - Discontinue
DISCD - Discontinued
DISCG - Discontinuing
DISRE - Disregard
DISRED - Disregarded
DISREG - Disregarding
DKTS - Dakotas
DLA - Delay
DLAD - Delayed
DLT - Delete
DLTD - Deleted
DLTG - Deleting
DLY - Daily
DMG - Damage
DMGD - Damaged
DMGG - Damaging
DMNT - Dominant
DMSH - Diminish
DMSHD - Diminished
DMSHG - Diminishing
DMSHS - Diminishes
DNS - Dense
DNSLP - Downslope
DNSTRM - Downstream
DNWND - Downwind
DP - Deep

DPND - Deepened
DPNG - Deepening
DPNS - Deepens
DPR - Deeper
DPTH - Depth
DRFT - Drift
DRFTD - Drifted
DRFTG - Drifting
DRFTS - Drifts
DRZL - Drizzle
DSCNT - Descent
DSIPT - Dissipate
DSIPTD - Dissipated
DSIPTG - Dissipating
DSIPTN - Dissipation
DSIPTS - Dissipates
DSND - Descend
DSNDG - Descending
DSNDS - Descends
DSNT - Distant
DSTBLZ - Destabilize
DSTBLZD - Destabilized
DSTBLZG - Destabilizing
DSTBLZS - Destabilizes
DSTBLZN - Destabilization
DSTC - Distance
DTRT - Deteriorate
DTRTD - Deteriorated
DTRTG - Deteriorating
DTRTS - Deteriorates
DURC - During climb
DURD - During descent
DURG - During
DURN - Duration
DVLP - Develop
DVLPD - Developed
DVLPG - Developing
DVLPMT - Development
DVLPS - Develops
DVRG - Diverge
DVRGG - Diverging
DVRGNC - Divergence
DVRGS - Diverges
DVV - Downward vertical velocity
DWNDFTS - Downdrafts
DWPNT - Dew point
DWPNTS - Dew points

E

E - East

EBND - Eastbound
EFCT - Effect
ELNGT - Elongate
ELNGTD - Elongated
ELSW - Elsewhere
EMBDD - Embedded
EMERG - Emergency
ENCTR - Encounter
ENDG - Ending
ENE - East-northeast
ENELY - East-northeasterly
ENERN - East-northeastern
ENEWD - East-northeastward
ENHNC - Enhance
ENHNCD - Enhanced
ENHNCG - Enhancing
ENHNCS - Enhances
ENHNCMNT - Enhancement
ENTR - Entire
ERN - Eastern
ERY - Early
ERYR - Earlier
ESE - East-southeast
ESELY - East-southeasterly
ESERN - East-southeastern
ESEWD - East-southeastward
ESNTL - Essential
ESTAB - Establish
EST - Estimate
ETA - Estimated time of arrival or ETA model
ETC - Et cetera
ETIM - Elapsed time
EVE - Evening
EWD - Eastward
EXCLV - Exclusive
EXCLVLY - Exclusively
EXCP - Except
EXPC - Expect
EXPCD - Expected
EXPCG - Expecting
EXTD - Extend
EXTDD - Extended
EXTDG - Extending
EXTDS - Extends
EXTN - Extension
EXTRAP - Extrapolate
EXTRAPD - Extrapolated
EXTRM - Extreme
EXTRMLY - Extremely
EXTSV - Extensive

F

F - Fahrenheit
FA - Aviation area forecast
FAM - Familiar
FCST - Forecast
FCSTD - Forecasted
FCSTG - Forecasting
FCSTR - Forecaster
FCSTS - Forecasts
FIG - Figure
FILG - Filling
FIR – Flight information region
FIRAV - First available
FL - Florida or flight level
FLG - Falling
FLRY - Flurry
FLRYS - Flurries
FLT - Flight
FLW - Follow
FLWG - Following
FM - From
FMT - Format
FNCTN - Function
FNT - Front
FNTL - Frontal
FNTS - Fronts
FNTGNS - Frontogenesis
FNTLYS - Frontolysis
FORNN - Forenoon
FPM - Feet per minute
FQT - Frequent
FQTLY - Frequently
FRM - Form
FRMG - Forming
FRMN - Formation
FROPA - Frontal passage
FROSFC - Frontal surface
FRST - Frost
FRWF - Forecast wind factor
FRZ - Freeze
FRZLVL - Freezing level
FRZN - Frozen
FRZG - Freezing
FT - Feet
FTHR - Further
FVRBL - Favorable
FWD - Forward
FYI - For your information

G

G - Gust
GA - Georgia
GEN - General
GENLY - Generally
GEO - Geographic
GEOREF - Geographical reference
GF - Fog
GICG - Glaze icing
GLFALSK - Gulf of Alaska
GLFCAL - Gulf of California
GLFMEX - Gulf of Mexico
GLFSTLAWR - Gulf of St. Lawrence
GND - Ground
GRAD - Gradient
GRDL - Gradual
GRDLY - Gradually
GRT - Great
GRTLY - Greatly
GRTLKS - Great Lakes
GSTS - Gusts
GSTY - Gusty
GTS – Global Telecommunication System

H

HAZ - Hazard
HDFRZ - Hard freeze
HDSVLY - Hudson Valley
HDWND - Head wind
HGT - Height
HI - High
HI - Hawaii
HIER - Higher
HIFOR - High level forecast
HLF - Half
HLTP - Hilltop
HLSTO - Hailstones
HND - Hundred
HPC – Hydrometeorological Prediction Center
HR - Hour
HRS - Hours
HRZN - Horizon
HTG - Heating
HURCN - Hurricane
HUREP - Hurricane report
HV - Have
HVY - Heavy
HVYR - Heavier
HVYST - Heaviest
HWVR - However

HWY - Highway

I

IA - Iowa
IC - Ice (in PIREPs only)
ICAO - International Civil Aviation
Organization
ICG - Icing
ICGIC - Icing in clouds
ICGICIP - Icing in clouds and in precipitation
ICGIP - Icing in precipitation
ID - Idaho
IFR - Instrument flight rules
IL - Illinois
IMDT - Immediate
IMDTLY - Immediately
IMPL - Impulse
IMPLS - Impulses
IMPT - Important
INCL - Include
INCLD - Included
INCLG - Including
INCLS - Includes
INCR - Increase
INCRD - Increased
INCRG - Increasing
INCRGLY - Increasingly
INCRS - Increases
INDC - Indicate
INDCD - Indicated
INDCG - Indicating
INDCS - Indicates
INDEF - Indefinite
INFO - Information
INLD - Inland
INSTBY - Instability
INTCNTL - Intercontinental
INTL - International
INTMD - Intermediate
INTMT - Intermittent
INTMTLY - Intermittently
INTR - Interior
INTRMTRGN - Intermountain region
INTS - Intense
INTSFCN - Intensification
INTSFY - Intensify
INTSFYD - Intensified
INTSFYG - Intensifying
INTSFYS - Intensifies
INTSTY - Intensity

INTVL - Interval
INVRN - Inversion
IOVC - In overcast
INVOF - In vicinity of
IP - Ice pellets
IPV - Improve
IPVG - Improving
ISOL - Isolate
ISOLD - Isolated

J

JCTN - Junction
JTSTR - Jet stream

K

KFRST - Killing frost
KLYR - Smoke layer aloft
KOCTY - Smoke over city
KS - Kansas
KT - Knots
KY - Kentucky

L

LA - Louisiana
LABRDR - Labrador
LAT - Latitude
LAWRS - Limited aviation weather reporting
        station
LCL - Local
LCLY - Locally
LCTD - Located
LCTN - Location
LCTMP - Little change in temperature
LEVEL - Level
LFTG - Lifting
LGRNG - Long-range
LGT - Light
LGTR - Lighter
LGWV - Long wave
LI - Lifted Index
LIS - Lifted Indices
LK - Lake
LKS - Lakes
LKLY - Likely
LLJ - Low level jet
LLWAS - Low-level wind shear alert system
LLWS - Low-level wind shear
LMTD - Limited

LMTG - Limiting
LMTS - Limits
LN - Line
LO - Low
LONG - Longitude
LONGL - Longitudinal
LRG - Large
LRGLY - Largely
LRGR - Larger
LRGST - Largest
LST - Local standard time
LTD - Limited
LTG - Lightning
LTGCC - Lightning cloud-to-cloud
LTGCG - Lightning cloud-to-ground
LTGCCCG - Lightning cloud-to-cloud cloud-to-ground
LTGCW - Lightning cloud-to-water
LTGIC - Lightning in cloud
LTL - Little
LTLCG - Little change
LTR - Later
LTST - Latest
LV - Leaving
LVL - Level
LVLS - Levels
LWR - Lower
LWRD - Lowered
LWRG - Lowering
LYR - Layer
LYRD - Layered
LYRS - Layers

M

MA - Massachusetts
MAN - Manitoba
MAX - Maximum
MB - Millibars
MCD - Mesoscale discussion
MD - Maryland
MDFY - Modify
MDFYD - Modified
MDFYG - Modifying
MDL - Model
MDLS - Models
MDT - Moderate
MDTLY - Moderately
ME - Maine
MED - Medium
MEGG - Merging

MESO - Mesoscale
MET - Meteorological
METAR - Aviation routine weather report
METRO - Metropolitan
MEX - Mexico
MHKVLY - Mohawk Valley
MI - Michigan
MID - Middle
MIDN - Midnight
MIL - Military
MIN - Minimum
MISG - Missing
MLTLVL - Melting level
MN - Minnesota
MNLD - Mainland
MNLY - Mainly
MO - Missouri
MOGR - Moderate or greater
MOV - Move
MOVD - Moved
MOVG - Moving
MOVMT - Movement
MOVS - Moves
MPH - Miles per hour
MRGL - Marginal
MRGLLY - Marginally
MRNG - Morning
MRTM - Maritime
MS - Mississippi
MSG - Message
MSL - Mean sea level
MST - Most
MSTLY - Mostly
MSTR - Moisture
MT - Montana
MTN - Mountain
MTNS - Mountains
MULT - Multiple
MULTILVL - Multilevel
MWO – Meteorological Watch Office
MXD - Mixed

N

N - North
NAB - Not above
NAT - North Atlantic
NATL - National
NAV - Navigation
NB - New Brunswick

NBND - Northbound
NBRHD - Neighborhood
NC - North Carolina
NCEP - National Center of Environmental
    Prediction
NCO – NCEP Central Operations
NCWX - No change in weather
ND - North Dakota
NE - Northeast
NEB - Nebraska
NEC - Necessary
NEG - Negative
NEGLY - Negatively
NELY - Northeasterly
NERN - Northeastern
NEWD - Northeastward
NEW ENG - New England
NFLD - Newfoundland
NGM - Nested grid model
NGT - Night
NH - New Hampshire
NIL - None
NJ - New Jersey
NL - No layers
NLT - Not later than
NLY - Northerly
NM - New Mexico
NMBR - Number
NMBRS - Numbers
NML - Normal
NMRS - Numerous
NNE - North-northeast
NNELY - North-northeasterly
NNERN - North-northeastern
NNEWD - North-northeastward
NNW - North-northwest
NNWLY - North-northwesterly
NNWRN - North-northwestern
NNWWD - North-northwestward
NNNN - End of message
NOAA - National Oceanic and Atmospheric
    Administration
NOPAC - Northern Pacific
NPRS - Nonpersistent
NR - Near
NRLY - Nearly
NRN - Northern
NRW - Narrow
NS - Nova Scotia
NTFY - Notify
NTFYD - Notified

NV - Nevada
NVA - Negative vorticity advection
NW - Northwest
NWD - Northward
NWLY - Northwesterly
NWRN - Northwestern
NWS - National Weather Service
NY - New York
NXT - Next

O

OAT - Outside air temperature
OBND - Outbound
OBS - Observation
OBSC - Obscure
OBSCD - Obscured
OBSCG - Obscuring
OCFNT - Occluded front
OCLD - Occlude
OCLDS - Occludes
OCLDD - Occluded
OCLDG - Occluding
OCLN - Occlusion
OCNL - Occasional
OCNLY - Occasionally
OCR - Occur
OCRD - Occurred
OCRG - Occurring
OCRS - Occurs
OFC - Office
OFP - Occluded frontal passage
OFSHR - Offshore
OH - Ohio
OK - Oklahoma
OMTNS - Over mountains
ONSHR - On shore
OR - Oregon
ORGPHC - Orographic
ORIG - Original
OSV - Ocean station vessel
OTLK - Outlook
OTP - On top
OTR - Other
OTRW - Otherwise
OUTFLO - Outflow
OVC - Overcast
OVHD - Overhead
OVNGT - Overnight
OVR - Over
OVRN - Overrun

OVRNG - Overrunning
OVTK - Overtake
OVTKG - Overtaking
OVTKS - Overtakes

P

PA - Pennsylvania
PAC - Pacific
PATWAS - Pilot's automatic telephone weather
      answering service
PBL - Planetary boundary layer
PCPN - Precipitation
PD - Period
PDMT - Predominant
PEN - Peninsula
PERM - Permanent
PGTSND - Puget Sound
PHYS - Physical
PIBAL - Pilot balloon observation
PIREP - Pilot weather report
PL – Ice pellets
PLNS - Plains
PLS - Please
PLTO - Plateau
PM - Postmeridian
PNHDL - Panhandle
POS - Positive
POSLY - Positively
PPINA - Radar weather report not available
PPINE - Radar weather report no echoes
      observed
PPSN - Present position
PRBL - Probable
PRBLY - Probably
PRBLTY - Probability
PRECD - Precede
PRECDD - Preceded
PRECDG - Preceding
PRECDS - Precedes
PRES - Pressure
PRESFR - Pressure falling rapidly
PRESRR - Pressure rising rapidly
PRIM - Primary
PRIN - Principal
PRIND - Present indications are
PRJMP - Pressure jump
PROB - Probability
PROC - Procedure
PROD - Produce
PRODG - Producing

PROG - Forecast
PROGD - Forecasted
PROGS - Forecasts
PRSNT - Present
PRSNTLY - Presently
PRST - Persist
PRSTS - Persists
PRSTNC - Persistence
PRSTNT - Persistent
PRVD - Provide
PRVDD - Provided
PRVDG - Providing
PRVDS - Provides
PS - Plus
PSBL - Possible
PSBLY - Possibly
PSBLTY - Possibility
PSG - Passage
PSN - Position
PSND - Positioned
PTCHY - Patchy
PTLY - Partly
PTNL - Potential
PTNLY - Potentially
PTNS - Portions
PUGET - Puget Sound
PVA - Positive vorticity advection
PVL - Prevail
PVLD - Prevailed
PVLG - Prevailing
PVLS - Prevails
PVLT - Prevalent
PWB - Pilot weather briefing
PWR - Power

Q

QN - Question
QSTNRY - Quasistationary
QTR - Quarter
QUAD - Quadrant
QUE - Quebec

R

R - Rain
RADAT - Radiosonde additional data
RAOB - Radiosonde observation
RCH - Reach

RCHD - Reached
RCHG - Reaching
RCHS - Reaches
RCKY - Rocky
RCKYS - Rockies
RCMD - Recommend
RCMDD - Recommended
RCMDG - Recommending
RCMDS - Recommends
RCRD - Record
RCRDS - Records
RCV - Receive
RCVD - Received
RCVG - Receiving
RCVS - Receives
RDC - Reduce
RDGG - Ridging
RDVLP - Redevelop
RDVLPG - Redeveloping
RDVLPMT - Redevelopment
RE - Regard
RECON - Reconnaissance
REF - Reference
RES - Reserve
REPL - Replace
REPLD - Replaced
REPLG - Replacing
REPLS - Replaces
REQ - Request
REQS - Requests
REQSTD - Requested
RESP - Response
RESTR - Restrict
RGD - Ragged
RGL - Regional model
RGLR - Regular
RGN - Region
RGNS - Regions
RGT - Right
RH - Relative humidity
RI - Rhode Island
RIOGD - Rio Grande
RLBL - Reliable
RLTV - Relative
RLTVLY - Relatively
RMK - Remark
RMN - Remain
RMND - Remained
RMNDR - Remainder
RMNG - Remaining
RMNS - Remains

RNFL - Rainfall
ROT - Rotate
ROTD - Rotated
ROTG - Rotating
ROTS - Rotates
RPD - Rapid
RPDLY - Rapidly
RPLC - Replace
RPLCD - Replaced
RPLCG - Replacing
RPLCS - Replaces
RPRT - Report
RPRTD - Reported
RPRTG - Reporting
RPRTS - Reports
RPT - Repeat
RPTG - Repeating
RPTS - Repeats
RQR - Require
RQRD - Required
RQRG - Requiring
RQRS - Requires
RSG - Rising
RSN - Reason
RSNG - Reasoning
RSNS - Reasons
RSTR - Restrict
RSTRD - Restricted
RSTRG - Restricting
RSTRS - Restricts
RTRN - Return
RTRND - Returned
RTRNG - Returning
RTRNS - Returns
RUF - Rough
RUFLY - Roughly
RVS - Revise
RVSD - Revised
RVSG - Revising
RVSS - Revises
RWY - Runway

S

S - South
SAB – Satellite Analysis Branch
SASK - Saskatchewan
SATFY - Satisfactory
SBND - Southbound
SBSD - Subside

SBSDD - Subsided
SBSDNC - Subsidence
SBSDS - Subsides
SC - South Carolina or stratocumulus
SCND - Second
SCNDRY - Secondary
SCSL -Stratocumulus standing lenticular
SCT - Scatter
SCTD - Scattered
SCTR - Sector
SD - South Dakota
SE - Southeast
SEC - Second
SELY - Southeasterly
SEPN - Separation
SEQ - Sequence
SERN - Southeastern
SEV - Severe
SEWD -Southeastward
SFC - Surface
SG - Snow grains
SGFNT - Significant
SGFNTLY - Significantly
SHFT - Shift
SHFTD - Shifted
SHFTG - Shifting
SHFTS - Shifts
SHLD - Shield
SHLW - Shallow
SHRT - Short
SHRTLY - Shortly
SHRTWV - Shortwave
SHUD - Should
SHWR - Shower
SIERNEV - Sierra Nevada
SIG - Signature
SIGMET - Significant meteorological
            information
SIMUL - Simultaneous
SKC - Sky clear
SKED - Schedule
SLD - Solid
SLGT - Slight
SLGTLY - Slightly
SLO - Slow
SLOLY - Slowly
SLOR - Slower
SLP - Slope
SLPG - Sloping
SLW - Slow
SLY - Southerly

SM - Statute mile
SML - Small
SMLR - Smaller
SMRY - Summary
SMTH - Smooth
SMTHR - Smoother
SMTHST - Smoothest
SMTM - Sometime
SMWHT - Somewhat
SN - Snow
SNBNK - Snowbank
SNFLK - Snowflake
SNGL - Single
SNOINCR - Snow increase
SNOINCRG - Snow increasing
SNST - Sunset
SOP - Standard operating procedure
SPC – Storm Prediction Center
SPCLY - Especially
SPD - Speed
SPKL - Sprinkle
SPLNS - Southern Plains
SPRD - Spread
SPRDG - Spreading
SPRDS - Spreads
SPRL - Spiral
SQ - Squall
SQLN - Squall line
SR - Sunrise
SRN - Southern
SRND - Surround
SRNDD - Surrounded
SRNDG - Surrounding
SRNDS - Surrounds
SS - Sunset
SSE - South-southeast
SSELY - South-southeasterly
SSERN - South-southeastern
SSEWD - South-southeastward
SSW - South-southwest
SSWLY - South-southwesterly
SSWRN - South-southwestern
SSWWD - South-southwestward
ST - Stratus
STAGN - Stagnation
STBL - Stable
STBLTY - Stability
STD - Standard
STDY - Steady
STFR - Stratus fractus
STFRM - Stratiform

STG - Strong
STGLY - Strongly
STGR - Stronger
STGST - Strongest
STM - Storm
STMS - Storms
STN - Station
STNRY - Stationary
SUB - Substitute
SUBTRPCL - Subtropical
SUF - Sufficient
SUFLY - Sufficiently
SUG - Suggest
SUGG - Suggesting
SUGS - Suggests
SUP - Supply
SUPG - Supplying
SUPR - Superior
SUPSD - Supersede
SUPSDG - Superseding
SUPSDS - Supersedes
SVG - Serving
SVRL - Several
SW - Southwest
SWD - Southward
SWWD - Southwestward
SWLY - Southwesterly
SWRN - Southwestern
SX - Stability index
SXN - Section
SYNOP - Synoptic
SYNS - Synopsis
SYS - System

T

TAF - Aviation terminal forecast
TCNTL - Transcontinental
TCU - Towering cumulus
TDA - Today
TEMP - Temperature
THK - Thick
THKNG - Thickening
THKNS - Thickness
THKR - Thicker
THKST - Thickest
THN - Thin
THNG - Thinning
THNR - Thinner
THNST - Thinnest
THR - Threshold

THRFTR - Thereafter
THRU - Through
THRUT - Throughout
THSD - Thousand
THTN - Threaten
THTND - Threatened
THTNG - Threatening
THTNS - Threatens
TIL - Until
TMPRY - Temporary
TMPRYLY - Temporarily
TMW - Tomorrow
TN - Tennessee
TNDCY - Tendency
TNDCYS - Tendencies
TNGT - Tonight
TNTV - Tentative
TNTVLY - Tentatively
TOPS - Tops
TOVC - Top of overcast
TPG - Topping
TRBL - Trouble
TRIB - Tributary
TRKG - Tracking
TRML -Terminal
TRMT - Terminate
TRMTD - Terminated
TRMTG - Terminating
TRMTS - Terminates
TRNSP - Transport
TRNSPG - Transporting
TROF - Trough
TROFS - Troughs
TROP - Tropopause
TRPCD - Tropical continental air mass
TRPCL - Tropical
TRRN - Terrain
TRSN - Transition
TS - Thunderstorm
TSFR - Transfer
TSFRD - Transferred
TSFRG - Transferring
TSFRS - Transfers
TSNT - Transient
TURBC - Turbulence
TURBT - Turbulent
TWD - Toward
TWDS - Towards
TWI - Twilight
TWRG - Towering
TX - Texas

U

UA - Pilot weather reports
UDDF - Up- and downdrafts
UN - Unable
UNAVBL - Unavailable
UNEC - Unnecessary
UNKN - Unknown
UNL - Unlimited
UNRELBL - Unreliable
UNRSTD - Unrestricted
UNSATFY - Unsatisfactory
UNSBL - Unseasonable
UNSTBL - Unstable
UNSTDY - Unsteady
UNSTL - Unsettle
UNSTLD - Unsettled
UNUSBL - Unusable
UPDFTS - Updrafts
UPR - Upper
UPSLP - Upslope
UPSTRM - Upstream
URG - Urgent
USBL - Usable
UT - Utah
UTC – Universal Time Coordinate
UVV - Upward vertical velocity
UWNDS - Upper winds

V

VA - Virginia
VAAC – Volcanic Ash Advisory Center
VAAS – Volcanic Ash Advisory Statement
VAL - Valley
VARN - Variation
VCNTY - Vicinity
VCOT - VFR conditions on top
VCTR - Vector
VFR - Visual flight rules
VFY - Verify
VFYD - Verified
VFYG - Verifying
VFYS - Verifies
VLCTY - Velocity
VLCTYS - Velocities
VLNT - Violent
VLNTLY - Violently

VMC - Visual meteorological conditions
VOL - Volume
VORT - Vorticity
VR - Veer
VRG - Veering
VRBL - Variable
VRISL - Vancouver Island, BC
VRS - Veers
VRT MOTN - Vertical motion
VRY - Very
VSB - Visible
VSBY - Visibility
VSBYDR - Visibility decreasing rapidly
VSBYIR - Visibility increasing rapidly
VT - Vermont
VV - Vertical velocity

W

W - West
WA - Washington
WAA - Warm air advection
WAFS – Word Area Forecast System
WBND - Westbound
WDLY - Widely
WDSPRD - Widespread
WEA - Weather
WFO - Weather Forecast Office
WFSO - Weather Forecast Service Office
WFP - Warm front passage
WI - Wisconsin
WIBIS - Will be issued
WINT - Winter
WK - Weak
WKDAY - Weekday
WKEND - Weekend
WKNG - Weakening
WKNS - Weakens
WKR - Weaker
WKST - Weakest
WKN - Weaken
WL - Will
WLY - Westerly
WND - Wind
WNDS - Winds
WNW - West-northwest
WNWLY - West-northwesterly
WNWRN - West-northwestern
WNWWD - West-northwestward
WO - Without
WPLTO - Western Plateau

WRM - Warm
WRMG - Warming
WRN - Western
WRMR - Warmer
WRMST - Warmest
WRMFNT - Warm front
WRMFNTL - Warm frontal
WRNG - Warning
WRS - Worse
WS - Wind shear
WSHFT - Windshift
WSFO - Weather Service Forecast Office
WSTCH - Wasatch Range
WSW - West-southwest
WSWLY - West-southwesterly
WSWRN - West-southwestern
WSWWD - West-southwestward
WTR - Water
WTSPT - Waterspout
WUD - Would
WV - West Virginia
WVS - Waves
WW - Severe weather watch
WWD - Westward
WX - Weather
WY - Wyoming

X

XCP - Except

XPC - Expect
XPCD - Expected
XPCG - Expecting
XPCS - Expects
XPLOS - Explosive
XTND - Extend
XTNDD - Extended
XTNDG - Extending
XTRM - Extreme
XTRMLY - Extremely

Y

YDA - Yesterday
YKN - Yukon
YLSTN - Yellowstone

Z

ZN - Zone
ZNS - Zones

SCHEDULED ISSUANCE AND VALID TIMES OF FORECAST PRODUCTS

Table 14-1 shows scheduled issuance and valid times of the TAFs. All times are UTC.

Table 14-1 Scheduled Issuance and Valid Times of TAFs

Scheduled Issuance Times	Valid Period	Transmission Period
00	00-00	2320-2340
06	06-06	0520-0540
12	12-12	1120-1140
18	18-18	1720-1740

The Table 14-2 has scheduled issuance and valid times of the TWEBs. All times are UTC.

Table 14-2 Scheduled Issuance and Valid Times of TWEBs

Scheduled Issuance Times	Valid Period	Transmission Period
02	02-14	0130-0140
08	08-20	0730-0740
14	14-02	1330-1340
20	20-08	1930-1940

Table 14-3 shows the scheduled issuance times of the FAs for their respective areas. The FA is valid 1 hour after issuance time. All times are UTC. The times the FA is issued depends on whether the FA area is in local standard or local daylight time.

Table 14-3 Scheduled Issuance Times of FAs

Area Forecast (FA)	Boston and Miami (LDT/LST)	Chicago and Ft. Worth (LDT/LST)	San Francisco and Salt Lake City (LDT/LST)	Alaska (LDT/LST)	Hawaii
1st issuance	0845/0945	0945/1045	1045/1145	0145/0245	0345
2nd issuance	1745/1845	1845/1945	1945/2045	0745/0845	0945
3rd issuance	0045/0145	0145/0245	0245/0345	1345/1445	1545
4th issuance				1945/2045	2145

Table 14-4 shows the scheduled issuance times of the Gulf of Mexico FA. All times are UTC.

Table 14-4 Scheduled Issuance Times of the Gulf of Mexico FA

Gulf of Mexico FA	Issuance Times (LDT/LST)
1st issuance	1040/1140
2nd issuance	1740/1840

NATIONAL WEATHER SERVICE STATION IDENTIFIERS

NORTHEAST REGION

AKQ - Norfolk/Wakefield, VA
ALY - Albany/East Berne, NY
BGM - Binghamton, NY
BOX - Boston/Taunton, MA
BTV - Burlington, VT
BUF - Buffalo, NY
CLE - Cleveland, OH
CTP - State College, PA
GYX - Portland/Gray, ME
ILN - Cincinnati/Wilmington, OH
LWX - Washington, DC/Sterling, VA
OKX - New York City/Brookhaven, NY
PBZ - Pittsburgh/Coraopolis, PA
PHI - Philadelphia, PA/Mount Holly, NJ
RLX - Charleston/Ruthdale, WV
RNK - Roanoke/Blacksburg, VA

SOUTHCENTRAL REGION

AMA - Amarillo, TX
BMX - Birmingham, AL
BRO - Brownsville, TX
CRP - Corpus Christi, TX
EPZ - El Paso, TX/Santa Theresa, NM
EWX - Austin/San Antonio, TX
FWD - Dallas/Forth Worth, TX
HGX - Houston/Dickinson, TX
JAN - Jackson, MS
LCH - Lake Charles, LA
LIX - New Orleans/Slidell, LA
LUB - Lubbock, TX
LZK - North Little Rock, AR
MAF - Midland, TX
MEG - Memphis/Germantown, TN
MOB - Mobile, MS
MRX - Knoxville/Tri Cities, TN
OHX - Nashville/Old Hickory, TN
OUN - Oklahoma City/Norman, OK
SHV - Shreveport, LA
SJT - San Angelo, TX
TSA - Tulsa, OK

SOUTHEAST REGION

CAE - Columbia, SC
CHS - Charleston, SC
FFC - Atlanta/Peachtree City, GA
GSP - Greenville-Spartanburg/Greer, SC
ILM - Wilmington, NC
JAX - Jacksonville, FL
MFL - Miami, FL
MHX - Morehead City/Newport, NC
MLB - Melbourne, FL
RAH - Raleigh/Durham, NC
TAE - Tallahassee, FL
TBW - Tampa/Ruskin, FL
TJSJ - San Juan, PR

MOUNTAIN REGION

ABQ - Albuquerque, NM
BIL - Billings, MT
BOI - Boise, ID
BOU - Denver/Boulder, CO
CYS - Cheyenne, WY
FGZ - Flagstaff/Bellemont, AZ
GGW - Glasgow, MT
GJT - Grand Junction, CO
LKN - Elko, NV
MSO - Missoula, MT
PIH - Pocatello, ID
PSR - Phoenix, AZ
PUB - Pueblo, CO
REV - Reno, NV
RIW - Riverton, WY
SLC - Salt Lake City, UT
TFX - Great Falls, MT
TWC - Tucson, AZ
VEF - Las Vegas, NV

NORTHCENTRAL REGION

ABR - Aberdeen, SD
APX - Alpena/Gaylord, MI
ARX - La Crosse, WI
BIS - Bismarck, ND
DDC - Dodge City, KS
DLH - Duluth, MN
DMX - Des Moines/Johnston, IA
DTX - Detroit/Pontiac, MI
DVN - Quad Cities/Davenport, IA

FGF - Fargo/Grand Forks, ND

PHFO - Honolulu, HI

EAX - Kansas City/Pleasant Hill, MO
FSD - Sioux Falls, SD
GID - Hastings, NE
GLD - Goodland, KS
GRB - Green Bay, WI
GRR - Grand Rpaids, MI
ICT - Wichita, KS
ILX - Lincoln, IL
IND - Indianapolis, IN
JKL - Jackson/Noctor, KY
LBF - North Platte, NE
LMK - Louisville, KY
LOT - Chicago/Romeoville, IL
LSX - St Louis, MO
MPX - Minneapolis/Chanhassen, MN
MKX - Milwaukee/Dousman, WI
MQT - Marquette, MI
OAX - Omaha/Valley, NE
PAH - Paducah, KY
SGF - Springfield, MO
TOP - Topeka, KS
UNR - Rapid City, SD

WEST COAST REGION

EKA - Eureka, CA
HNX - Hanford, CA
LOX - Los Angeles/Oxnard, CA
MFR - Medford, OR
MTR - San Francisco/Monterey, CA
OTX - Spokane, WA
PDT - Pendelton, OR
PQR - Portland, OR
SEW - Seattle, WA
SGX - San Diego, CA
STO - Sacramento, CA

ALASKAN REGION

PAFC - Anchorage, AK
PAFG - Fairbanks, AK
PAJK - Juneau, AK

PACIFIC REGION

PGUA - Tiyan, GU

WSR-88D SITES

ABC Bethel, AK
ABR Aberdeen, SD
ABX Albuquerque, NM
ACG Sitka/Biorka Island, AK
AEC Nome, AK
AHG Anchorage/Nikiski, AK
AIH Middleton Island, AK
AKC King Salmon, AK
AKQ Norfolk/Wakefield, VA
AMA Amarillo, TX
AMX Miami, FL
APD Fairbanks, AK
APX Gaylord, MI
ARX La Crosse, WI
ATX Seattle-Tacoma/Camano Island, WA
BBX Marysville/Beale AFB, CA
BGM Binghamton, NY
BHX Eureka/Bunker Hill, CA
BIS Bismarck, ND
BIX Keesler AFB, MS
BLX Billings/Yellowstone County, MT
BMX Birmingham/Alabaster, AL
BOX Boston/Taunton, MA
BRO Brownsville, TX
BUF Buffalo/Cheektowaga, NY
BYX Key West/Boca Chica Key, FL
CAE Columbia, SC
CBW Caribou/Hodgdon, ME
CBX Boise/Ada County, ID
CCX State College/Rush, PA
CLE Cleveland, OH
CLX Charleston/Grays, SC
CRP Corpus Christi, TX
CXX Burlington/Colchester, VT
CYS Cheyenne, WY
DAX Sacramento, CA
DDC Dodge City, KS
DFX Del Rio/Laughlin AFB, TX
DIX Philadelphia, PA/Fort Dix, NJ
DLH Duluth, MN
DMX Des Moines/Johnston, IA
DOX Dover AFB , DE
DTX Detroit-Pontiac/White Lake, MI
DVN Quad Cities/Davenport, IA
DYX Abilene/Dyess AFB, TX
EAX Kansas City/Pleasant Hill, MO
EMX Tucson/Pima County, AZ
ENX Albany/East Berne, NY
EOX Fort Rucker, AL

EPZ El Paso, TX/Santa Teresa, NM
ESX Las Vegas/Nelson, NV
EVX Red Bay/Eglin AFB, FL
EWX Austin-San Antonio/New Braunfels, TX
EYX Edwards AFB, CA
FCX Roanoke/Coles Knob, VA
FDR Frederick/Altus AFB, OK
FDX Clovis/Cannon AFB, NM
FFC Atlanta/Peachtree City, GA
FSD Sioux Falls, SD
FSX Flagstaff/Coconino, AZ
FTG Denver/Boulder, CO
FWS Dallas/Fort Worth, TX
GGW Glasgow, MT
GJX Grand Junction/Mesa, CO
GLD Goodland, KS
GRB Green Bay/Ashwaubenon, WI
GRK Killeen/Fort hood, TX
GRR Grand Rapids, MI
GSP Greenville-Spartanburg/Greer, SC
GUA Agana, GU
GWX Columbus AFB, MS
GYX Portland/Gray, ME
HDX Alamogordo/Holloman AFB, NM
HGX Houston-Galveston/Dickinson, TX
HKI South Kauai/Numila, HI
HKM Kamuela/Puu Mala, HI
HMO Molokai/Kukui, HI
HNX San Joaquin Valley/Hanford, CA
HPX Fort Campbell, KY
HTX Hytop, AL
HWA South Hawaii/Naalehu, HI
ICT Wichita, KS
ICX Cedar City, UT
ILN Cincinnati/Wilmington, OH
ILX Lincoln, IL
IND Indianapolis, IN
INX Tulsa/Inola, OK
IWA Phoenix/Mesa, AZ
IWX North Webster, IN
JAN Jackson, MS
JAX Jacksonville, FL
JGX Warner Robins/Robins AFB, GA
JKL Jackson/Noctor, KY
JUA San Juan/Cayey, PR
LBB LUBBOCK, TX
LCH Lake Charles, LA
LIX New Orleans-Baton Rouge/Slidell, LA
LNX North Platte/Thedford, NE
LOT Chicago/Romeoville, IL
LRX Elko/Sheep Creek Mountain, NV

LSX ST. Louis/Research Park, MO
LTX Wilmington/Shallotte, NC
LVX Louisville/Fort Knox, KY
LWX Baltimore, MD-Washington,
      DC/Sterling, VA
LZK North Little Rock, AR
MAF Midland/Odessa, TX
MAX Medford/Mount Ashland, OR
MBX Minot AFB, ND
MHX Morehead City/Newport, NC
MKX Milwaukee/Dousman, WI
MLB Melbourne, FL
MOB Mobile, AL
MPX Minneapolis/Chanhassen, MN
MQT Marquette/Negaunee, MI
MRX Knoxville-Cities/Morristown, TN
MSX Missoula/Point Six Mountain, MT
MTX Salt Lake City/Promontory Point, UT
MUX San Francisco/Mount Umunhum, CA
MVX Fargo-Grand Forks/Mayville, ND
MXX Carrville/Maxwell AFB, AL
NKX San Diego/Miramar Nas, CA
NQA Memphis/Millington, TN
OAX Omaha/Valley, NE
OHX Nashville/Old Hickory, TN
OKX New York City/Upton, NY
OTX Spokane, WA
PAH Paducah, KY
PBZ Pittsburgh/Coraopolis, PA
PDT Pendleton, OR

POE Fort Polk, LA
PUX Pueblo, CO
RAX Raleigh-Durham/Clayton, NC
RGX Reno/Virginia Peak, NV
RIW Riverton, WY
RLX Charleston/Ruthdale, WV
RMX Rome/Griffiss AFB, NY
RTX Portland/Scappoose, OR
SFX Pocatello-Idaho Falls/Springfield, ID
SGF Springfield, MO
SHV Shreveport, LA
SJT San Angelo, TX
SOX Santa Ana Mountains/Orange County, CA
SRX Slatington Mountain, AR
TBW Tampa/Ruskin, FL
TFX Great Falls, MT
TLH Tallahassee, FL
TLX Oklahoma City/Norman, OK
TWX Topeka/Alma, KS
TYX Fort Drum, NY
UDX Rapid City/New Underwood, SD
UEX Hastings/Blue Hill, NE
VAX Valdosta/Moody AFB, GA
VBX Lompoc/Vandenberg AFB, CA
VNX Enid/Vance AFB, OK
VTX Los Angeles/Sulphur Mountain, CA
YUX Yuma, AZ

INTERNET ADDRESSES

NATIONAL WEATHER SERVICE HOME PAGE

http://www.nws.noaa.gov

INTERACTIVE WEATHER INFORMATION NETWORK (IWIN)

http://weather.gov

WEATHER CHARTS

http://weather.noaa.gov/fax/graph.shtml
or
http://weather.noaa.gov/fax/nwsfax.shtml

AVIATION DIGITAL DATA SERVICE

http://adds.awc-kc.noaa.gov

NWS NATIONAL CENTERS FOR ENVIRONMENTAL PREDICTION

http://www.ncep.noaa.gov

AVIATION WEATHER CENTER

http://www.awc-kc.noaa.gov

NWS LINKS

http://nimbo.wrh.noaa.gov/wrhq/nwspage.html
or
http://www.nws.noaa.gov/regions.shtml

ALASKAN AVIATION WEATHER UNIT

http://www.alaska.net/~aawu/